The decades preceding the First World War constitute an era in which it is often claimed that the origins of Britain's relative economic decline are first witnessed. For the papermaking industry this was also a period in which an array of important new forces, including *inter alia* the development of new raw materials and the move to ever larger scales of production, came on the scene. Gary Bryan Magee examines the effect of these changes and assesses how effectively the industry coped with the new pressures, drawing upon an extensive range of quantitative and archival sources from Britain, America, and other countries. Along the way, Dr Magee addresses significant economic issues central to the understanding of industrial competitiveness, such as technological change, entrepreneurship, productivity, trade policy, and industrial relations.

Cambridge Studies in Modern Economic History 4

Productivity and performance
in the paper industry

Cambridge Studies in Modern Economic History

Series editors

Charles Feinstein
All Souls College, Oxford

Patrick O'Brien
The Institute of Historical Research, London

Barry Supple
The Leverhulme Trust

Peter Temin
Massachusetts Institute of Technology

Gianni Toniolo
Universita degli Studi di Venezia

Cambridge Studies in Modern Economic History is a major new initiative in economic history publishing, and a flagship series for Cambridge University Press in an area of scholarly activity in which it has long been active. Books in this series will be concerned primarily with the history of economic performance, output and productivity, assessing the characteristics, causes and consequences of economic growth (and stagnation) in the western world. This range of enquiry, rather than any other methodological or analytic approach, will be the defining characteristic of volumes in the series.

The first titles in the series are:
1 *Central and Eastern Europe 1944–1993: detour from the periphery to the periphery*
Ivan Berend
ISBN 0 521 55066 1

2 *Spanish agriculture: the long Siesta 1765–1965*
James Simpson
ISBN 0 521 49030 6

3 *Democratic socialism and economic policy: the Attlee years 1945–1951*
Jim Tomlinson
ISBN 0 521 55095 5

4 *Productivity and performance in the paper industry: labour, capital, and technology in Britain and America, 1860–1914*
Gary Bryan Magee
ISBN 0 521 58197 4

Productivity and performance in the paper industry

Labour, capital, and technology in Britain and America, 1860–1914

Gary Bryan Magee

Institute of Advanced Studies
The Australian National University

PUBLISHED BY THE PRESS SYNDICATE OF THE UNIVERSITY OF CAMBRIDGE
The Pitt Building, Trumpington Street, Cambridge, United Kingdom

CAMBRIDGE UNIVERSITY PRESS
The Edinburgh Building, Cambridge CB2 2RU, UK
40 West 20th Street, New York NY 10011-4211, USA
477 Williamstown Road, Port Melbourne, VIC 3207, Australia
Ruiz de Alarcón 13, 28014 Madrid, Spain
Dock House, The Waterfront, Cape Town 8001, South Africa

http://www.cambridge.org

First published 1997
First paperback edition 2002

A catalogue record for this book is available from the British Library

Library of Congress Cataloguing in Publication data
Magee, Gary Bryan, 1965–
Productivity and performance in the paper industry: labour, capital, and
technology in Britain and America, 1860–1914 / Gary Bryan Magee.
        p.   cm. – (Cambridge studies in modern economic history: 4)
Includes bibliographical references.
ISBN 0 521 58197 4 (hardcover)
1. Paper industry – Great Britain – History – 19th century.
2. Paper industry – Great Britain – History – 20th century.
3. Paper industry – United States – History – 19th century.
4. Paper industry – United States – History – 20th century.
5. Industrial productivity – Great Britain – History – 19th century.
6. Industrial productivity – Great Britain – History – 20th century.
7. Industrial productivity – United States – History – 19th century.
8. Industrial productivity – United States – History – 20th century.
I. Title. II. Series.
HD9831.5.M34   1997
338.4'76762'0941–dc20   95-26082   CIP

ISBN 0 521 58197 4 hardback
ISBN 0 521 89217 1 paperback

For my parents, Robert and Christina Magee,
and my wife, Min Mee

Contents

Figures

Tables

Acknowledgements

This work has benefited from the advice and comments of many friends and colleagues, including James Foreman-Peck, Avner Offer, Charles Feinstein, Graeme Snooks, Christine MacLeod, Sue Bowden, and Steve Broadberry. Useful insights were also provided by participants at seminars in Oxford, Canberra and Hagley and at the 1992 Economic History Society Conference in Hull. All remaining errors, of course, are my own.

This book developed out of my D.Phil. thesis at Nuffield College, Oxford. Funding for my doctoral research was provided by the Commonwealth Scholarship Commission, while generous financial support for a research trip to the United States was also given by Hagley Museum and Library, the Sir John Hicks Fund and the Goodhardt Trust. A postdoctoral fellowship at the Australian National University provided the opportunity and time to convert my doctoral thesis into this book.

Over the past three and a half years, I have received a great deal of help from librarians and archivists, too numerous to mention individually, in Britain, America, Australia and Austria. My research has also been aided by Bowater plc, which kindly granted me access to their archives as well as to those of Edward Lloyd Limited, which the company still retains. In particular, I would like to express my gratitude to Bowater's archivist, Mr H. J. Kennard, for his invaluable and friendly assistance in helping me track down all of the relevant documents that I needed to see. My thanks also go to Wayne Naughton for his excellent technical assistance.

Finally, I would like to thank my wife, Min Mee, and my parents for their unfailing understanding and support.

Abbreviations

AFL	American Federation of Labor
Amalgamated	Amalgamated Society of Paper Makers
APPA	American Paper and Pulp Association
Bertrams Collection	James Bertram and Son Limited Records
BLPES	British Library of Political and Economic Science
Brown Collection	Messrs Brown and Company Limited Records
CKS	Centre for Kentish Studies
Correspondence	Copy of Correspondence between Departments of Treasury and Board of Trade, in regard to the increasing Scarcity of the materials for the Fabrication of Paper (1854)
Cowans Collection	Alexander Cowan and Sons Limited Records
Earnings and Hours	Report of an Enquiry by the Board of Trade into the Earnings and Hours of Labour of Workpeople of the United Kingdom (1912/13)
Fourth Report on Employment	Fourth Report of the Commissioners on the Employment of Children and Young Persons in Trades and Manufactures not already regulated by Law
HML	Hagley Museum and Library
IBPM	International Brotherhood of Paper Makers
IBPSPMW	International Brotherhood of Pulp, Sulphite, and Paper Mill Workers
IPC	International Paper Company

IPMTU	International Paper Machine Tenders Union
MRC	Modern Records Centre
NC	Nuffield College
NCF	National Civic Federation
NICB	National Industrial Conference Board
NPV	Net Present Value
NUPMW	National Union of Paper Mill Workers
OCRO	Oxfordshire County Records Office
OSP	Original Society of Paper Makers
Paper Mill Directory	*Paper Mill Directory of England, Scotland, and Ireland and Year Book of the Paper Making Trade*
PMBPTJ	*Paper Maker and British Paper Trade Journal*
PMC	*Paper Makers' Circular and Rag Merchant Gazette and Price Current*
PP	*British Parliamentary Papers*
PTJ	*Paper Trade Journal*
PTR	*Paper Trade Review*
Pulp and paper	US *Pulp and Paper Investigation Hearings* (1909)
RCA	Revealed Comparative Advantage
RC Depression	Royal Commission Appointed to Inquire into the Depression of Trade and Industry (1886)
RC Labour	Royal Commission on Labour – Textiles, Clothing, Chemical, Buildings and Miscellaneous Trades (Group C) (1893/4)
SC Paper	Report from the Select Committee on Paper (Export Duty on Rags) (1861)
Social Reconstruction Survey	Nuffield College Social Reconstruction Survey
SRO	Scottish Records Office
Tariff Collection	Tariff Commission Collection
TFP	Total Factor Productivity
Webb Collection	Webb Trade Union Collection
Verband	Verband deutscher Druckpapierfabriken
Verein	Verein deutscher Papierfabrikanten

Introduction

Nothing attracts the attention of historians, economists, and even the more astute of policy-makers, like the economic decline of once great and powerful societies. This is amply borne out by the plethora of theories accounting for the demise of *inter alia* Ancient Rome, Venice, Holland, Imperial China, and, in more recent times, America and the West in general. It is, moreover, an observation that seems to hold *a fortiori* when the declining society had once been the world's most dynamic. It is in this light that post-1870 Britain has come to be regarded by many.

That this should be so is hardly surprising. In the hundred years prior to the 1870s Britain had managed to break from the pack of rival European nations to command, what must have appeared to contemporaries, a seemingly unassailable lead in industrial production. The events of the nineteenth century, however, soon revealed that Britain's historic role was to be that of *primus inter pares*, not workshop of the world. By the turn of the century Britain was thus only one of several industrialised countries, albeit an important one. Increasingly British firms found themselves driven by foreign competitors from markets that they had once pioneered and dominated.

In many ways this was the experience of the British paper industry in the nineteenth century. From a situation at the beginning of the century where British papermakers were amongst the most advanced producers of paper in the world – a lead primarily won by their early introduction of the Fourdrinier paper-machine – the industry in the latter half of the century seemed to enter something of a relative decline, characterised by growing foreign import penetration and shrinking market shares. By the 1860s and 1870s Britain's chief competitors had likewise mechanised the production of paper in their countries and made great strides in the search for, and development of, new raw materials.

Yet, in a number of interesting respects the paper industry differed from the old, often moribund, industries that had already begun to fade in Britain in the late Victorian and Edwardian era. Although historically an old industry, by the late nineteenth century the paper industry in Britain had taken on a remarkably modern guise. Indeed, dominated by a capital-intensive flow production technology and a process of innovation characterised by gradual technological accumulation rather than major leaps, the paper industry rather than being a mere relic of the past was in many ways more a foretaste of the future. Moreover, with its long, standardised production runs and high proportion of relatively unskilled labourers working on daily wages, the industry stands as a useful balance to the widely held conception that all British industry in the nineteenth century was geared to the batch production of goods by highly skilled craft labour on piece-rates. That such a 'modern' industry should in any case decline just as emphatically from the second half of the nineteenth century as the old staple industries did, should in itself arouse interest in the industry, raising as it does questions central to the cause and nature of Britain's relative economic decline in general since the nineteenth century.

The British paper industry, however, has hitherto been strangely neglected by economic historians. Although Spicer undoubtedly claims too much in saying that the advent of cheap paper had a far greater impact on British society than the steam engine, it is nonetheless surprising just how little attention the industry has actually attracted.[1] Conceivably, the smallness of the industry's contribution to the overall British manufacturing sector has had something to do with the relatively poor coverage it has received, yet with just under 5 per cent of the labour force in manufacturing finding employment within its ranks in 1891, the paper industry was in this respect actually larger than both the chemical and the food, drink, and tobacco industries. Of course, the practice of equating an industry's importance with its size is a dangerous and unreliable one to adopt.[2] This is especially so in this instance, as the role paper has played, and still plays, in sustaining modern society certainly far outweighs the industry's perceived deficiencies in scale. As a consequence of this role paper has become a ubiquitous part of modern society; a product without which life would be far more complicated and inconvenient.

The aim of this book is to assess and explain the performance of the British paper industry in the late Victorian and Edwardian period. Along

[1]   A. D. Spicer, *The Paper Trade* (London, 1907), p. 2.
[2]   It should be remembered that in 1900 cotton employed only just over 3 per cent of the British labour force. Smaller than the paper industry were the chemical and the food, drinks, and tobacco industries, whose workforces in 1891 made up just 1.7 and 4.1 per cent of all labourers employed in manufacturing. The corresponding figure for the paper industry in that year was 4.9 per cent. P. Deane and W. A. Cole, *British Economic Growth* (Cambridge, 1967), p. 146.

the way it necessarily explores concepts and processes such as technological change, productivity growth, free trade, and entrepreneurship, that lie at the very heart of industrial competitiveness. Despite their proximity to the heart of the matter, however, many of these issues and phenomena remain to a large extent empty boxes; plainly important, but largely taken for granted. Entrepreneurship is a case in hand. Economic historians, who engage in the debate on entrepreneurial failure in late Victorian and Edwardian Britain, usually delve no deeper into the dynamics of entrepreneurial activity and decision-making than the neoclassical or Schumpeterian traditions allow. With the neoclassical approach, that amounts to the equating of entrepreneurship with management, and hence, in effect, the assuming away of the entrepreneur. In this case entrepreneurship becomes indistinguishable from profit-maximisation. By contrast, in the Schumpeterian schema the distinctive activities of the entrepreneur are lauded and elevated to a central position. Nevertheless, although vital to the dynamics of the whole system, this tradition still regards entrepreneurship as being exogenously determined.

At least as practised in the economic history literature then, there would appear to be no adequate theory for what it is the entrepreneur does, and how he or she goes about doing it. Since it is the entrepreneur who is often asked to accept the blame for the failings of the British economy from the last quarter of the nineteenth century, this seems a rather strange omission. It is surely valid to ask how one can be certain that entrepreneurship is indeed at fault when no attempt is made even to define it, let alone to provide guidelines by which its quality can be assessed. To do that, of course, would require one to know something about how entrepreneurs make decisions; or in other words, to understand at a micro level how he or she works.

This book posits such a model of entrepreneurial decision-making. The model is based on the notion that as most decision-making is of an on-going nature, information – the bedrock of all decision-making – and consequentially decision-rules must be in a state of constant flux. When assessing decisions it is therefore necessary to consider the quantity and quality of information available at the time the decision is made. In the early stages, when information is patchy and incomplete, intuition and hunch are central to the decision-making process. However, as the entrepreneur's data base expands – often in the direction laid down by his or her own previous decisions and choices – more rationally based decisions become feasible. In time, when the various options have been more fully worked out and understood, the type of optimisation behaviour described by neoclassical economics becomes a possibility. As very little entrepreneurial decision-making actually takes place in such a setting, however, the best that one can ever reasonably expect from one's entrepreneurs is for them to act with competence and vigour.

In chapter 4 this framework is outlined and applied to the papermaker's search for new raw materials in the second half of the century: one of the most dramatic and testing events in the industry's history. It was a quest which in the end turned out to be long and arduous and which not only questioned the very strength and quality of British entrepreneurship, but also saw Britain for some thirty years pursue an entirely different path from its main competitor in the trade, America. Since British entrepreneurs have frequently been criticised for their failure to adopt American best-practice, and as there is also little doubt that the decision directly impinged on the industry's ability to perform, the search provides an ideal test case for the entrepreneurial failure thesis in the late Victorian paper industry.

Intimately tied to entrepreneurship and just as little understood in the entrepreneurial failure literature is technological change. Frequently depicted as manna from heaven and as homogeneous in character, its occurrence in that literature is more often assumed than explained. In reality, the process is much more diverse and varied. After all, there is no real reason for us to believe that all technological change should be similar in either pattern, direction, or timing. While there are general trends, whose influences will be fairly universally felt, the process in each industry is more likely to be affected by conditions that are unique to that industry and its technology. As a consequence, the variety of types and patterns of technological change possible is wide.

In nineteenth-century papermaking, technological progress was principally derived from extensions and improvements to the existing technology made possible by the learning and accumulation of knowledge attained through the act of production. By nature, this was an incremental, cumulative, and often firm-specific process. The effects of this type of learning on technological change, however, are not well understood in the theoretical literature. In particular, little attention in economics and economic history has been devoted to the determinants of such economies of practice. As a start, chapter 2 of this book explores some of the factors that could influence the degree of innovation achieved via learning; an analysis clearly of use to all industries where such forms of technological change are important. It concludes that if one is to understand the rate of technological accumulation in an industry, one must look at factors that affect that industry's opportunity, ability, and willingness to learn from production. Such analysis would involve consideration of a wide range of factors, ranging *inter alia* from institutional and organisational setting to the development of human capital and labour relations. In chapter 8 this approach is applied to the Anglo-American technological divide that appeared in the 1890s.

A number of issues and phenomena common to late Victorian Britain and beyond also surface in the paper industry at this time. One such facet of the industry's experience, like that of the manufacturing sector as a whole, was the American industry's higher level of labour productivity from 1860 when Britain was still the technological leader in the trade. This curious feature raises interesting questions about the British industry's performance in the second half of the nineteenth century which, if answered, may not only aid our understanding of the relative decline of British papermaking, but also contribute to our knowledge of the determinants of technological and productivity leadership in general: factors crucial to the economic rise and fall of nations. In particular, this book considers in chapters 5 through to 8 to what extent this productivity gap between Britain and America reflected each country's own distinct resource environment and pattern of demand, and to what extent the gap stemmed from other factors such as rates of technological change and the actions and attitudes of each country's workforce and entrepreneurs.

Also of general interest is the impact national commercial policies had on the industry's progress. In October 1861 Britain removed the last of its import duties on paper and board, for the first time exposing its manufacturers to the full force of competition from foreign producers who themselves continued to be protected by imposing tariff walls. The paper industry thus affords an opportunity to examine how this combination of British free trade and foreign protectionism impacted on an industry like the paper industry where long production runs and economies of scale were vital to survival. In such circumstances it is argued that the effect was far from negligible. In chapter 9 British and German trade in paper in particular is examined from this perspective.

Readers will find the methodology adopted in this book somewhat eclectic, as an amalgam of quantitative, qualitative, and theoretical evidence is employed. The absence of any single comprehensive source on the industry makes such an approach essential. To this end, a wide variety of primary sources, including patent data, government publications, trade journals, contemporary manuscripts, private notes, union records, and business archives are used in conjunction with secondary literature in an attempt to make some sort sense out of the many intertwining trends and forces operating in the latter half of the nineteenth century. Direct and detailed comparisons with the British industry's leading competitors – America and Germany – are also consciously made in this book in the belief that such comparisons are not only helpful in putting the British industry's performance into its proper international context, but also serve to elucidate many of the crucial aspects of the industry's development in the late Victorian and Edwardian era.

1   Background

Viewed from an international perspective, papermaking in Britain is a relatively recent phenomenon. Although the ability to make paper had filtered through from China via the Middle East to continental Europe by the eleventh century, such knowledge appears not to have been absorbed, or at least implemented, in the British Isles until the end of the fifteenth century when John Tate began operating his mill on the River Lee near Hertford. Within twenty years of its opening, however, this mill, like so many other early paper-mills, had ceased production, and for the next two centuries the industry's existence on these shores remained precarious. Despite its relatively late arrival on the scene, however, the British industry has nonetheless managed to make a disproportionately large and lasting contribution to the trade's economic and technological development. Amongst its achievements can be listed *inter alia* the first successful mechanisation of the production process, the development of new raw materials such as esparto grass, and the introduction of a wide range of new paper products such as security and bible paper. Yet it is a list of accomplishments that has been largely forgotten by all except a handful of devoted paper historians. Indeed, the number of scholarly works written about the industry's history in Britain can be counted on one hand, with economic history's contribution to this total being virtually limited to A. D. Spicer's *The Paper Trade* (1907), Donald Coleman's *The British Paper Industry, 1495–1860* (1958), and Richard Hills' *Papermaking in Britain, 1488–1988* (1988). Large and important areas of the industry's history, therefore, have been left largely unexplored. One of these is the late Victorian and Edwardian period; a period in which an array of important new forces, all of which in their own way were to transform the industry irrevocably, came on the scene. In essence, this book is about how effectively the industry coped with these new pressures. It is a book among other things about technological change, institutions, and entrepreneurship, and how these impacted on performance. Given the nature of many of the issues addressed, it is also a book that perforce has a strong international dimension.

Before we can turn our attention to such questions of performance, however, some of the more fundamental characteristics of the paper trade need to be examined. The purpose of this, and the following chapter on technological change, is thus to highlight some of the more important supply- and demand-side features of the late Victorian and Edwardian paper industry. Such preliminary information provides a background to the analyses of later chapters. In this chapter attention is chiefly focussed on the industry's location, cost structure, market structure, and demand.[1]

Location

In composition, paper is simply a mixture of cellulose and water (called pulp) that has been dried into the form we are familiar with by the joint effects of evaporation, gravity, and artificial heating. By the middle of the nineteenth century, this process, although originally performed by hand, had become largely a mechanical one dominated by a single machine, the Fourdrinier. Invented in Paris in 1799 by Nicholas Louis Robert, and first introduced into Britain by the Fourdrinier brothers in their Frogmore Mill in Hertfordshire in 1804, this remarkable machine, once fully worked out, was able, in one continuous process, to transform pulp into finished paper. Given the indispensable position in the trade the paper-machine quickly assumed for itself in the following half century, matters concerning the paper-machine understandably came to exert a significant influence over the structure and character of papermaking in Britain. In this regard, however, it was not alone. The raw materials needed to manufacture paper also had an important role to play in determining the nature and shape of the industry that emerged in the nineteenth century. The basis of all paper, cellulose, is a natural fibre found in most vegetation, as well as in items, such as cotton or linen rags, which are essentially made from some form of plant extract. The other vital input in the process, of course, was water. Indeed, so significant was this consideration that, since the industry's inception in the fifteenth century, the distribution of paper-mills in Britain has for most of this time been governed primarily by its availability; a condition that stemmed from the fact that water was not only a crucial ingredient in the production process, but prior to the steam engine was also the

[1] Throughout this book the paper industry is regarded as a single entity. While it is true that the paper industry, like many others, can be broken down into a number of separate branches, each making distinct products, the homogeneity of conditions and production technology prevailing in the trade allows this assumption of unity to be reasonably made. Of course, wherever a particular branch of the trade varies from this norm in some respect, appropriate note of these differences will be given.

Background
9

industry's only source of energy.[2] The other distinctive feature of the early industry's location was its development near large towns and cities that could offer good markets and supplies of papermaking materials. The most important of these raw materials has traditionally been rag. Prior to the eighteenth century, such criteria acted together to create an industry predominantly located in the south of the country, especially in the Home Counties. Kent's quick streams of clear, hard water derived from contact with the region's limestone, not to mention its close proximity to London and its rich supply of rags, made it an especially suitable base for the industry. Mills also cropped up in the Mendips while the old cordage and sail widely available in ports such as Exeter, Dover, and Southampton gave rise to early brown and wrapping paper industries in these cities.[3]

Papermaking, however, was not to remain forever tied to the south of England. Over the centuries the industry began to spread from its traditional base in a fashion very much in line with the nation's population and industrial growth. As a result one finds that by the end of the eighteenth century it was already established in all counties of England, Scotland, and Wales, and that Lancashire, Yorkshire, and the Edinburgh region were now beginning to rival the former pre-eminence of the home counties. From the early decades of the nineteenth century the advent and diffusion of steam-powered paper-machines markedly accelerated this geographical spread of the industry by providing the means to liberate it from its traditional dependence on fast-running water.[4]

Using information contained in paper trade directories, snapshots of the changing regional distribution of mills over the second half of the nineteenth century can be captured. As these show, by 1865 all regions but East Anglia contained at least ten mills within their boundaries. The importance of the South East, North, and Scotland is also immediately clear. In 1865 these three regions together comprised some 69.6 per cent of all mills in the country. In that year the South West and the Midlands were also home to large numbers of papermakers. By 1914, however, this relatively even distribution of mills had already eroded to the point where the South West, East Anglia, and the East Midlands were no longer major loci of the industry. The decline in these regions was, in turn, matched by a greater proportion of the industry finding itself in the West Midlands as well as in the traditional centres of the

[2] The process of making paper is described in more detail in chapter 2.
[3] D. C. Coleman, *The British Paper Industry, 1495–1860: A Study in Industrial Growth* (Oxford, 1958), pp. 34–8.
[4] *Ibid.*, pp. 146–9.

Table 1.1 *Distribution of mills by regions, 1865 and 1914 (percentages in brackets)*

Region	1865	1914
South East	96 (27.7)	73 (24.9)
South West	34 (9.8)	18 (24.9)
East Anglia	7 (2.0)	2 (0.7)
East Midlands	34 (9.8)	19 (6.5)
West Midlands	20 (5.8)	25 (8.5)
Wales	10 (2.9)	7 (2.4)
North	91 (26.3)	87 (29.7)
Scotland	54 (15.6)	62 (21.2)

Notes: South East – Hampshire, Sussex, Kent, Essex, Hertford, Buckinghamshire, Berkshire, Middlesex, Surrey, London; South West – Cornwall, Devon, Somerset, Dorset, Wiltshire; East Anglia – Cambridgeshire, Norfolk, Suffolk; East Midlands – Derbyshire, Nottinghamshire, Lincolnshire, Warwickshire, Leicestershire, Rutland, Northamptonshire, Huntingdon, Oxfordshire, Bedfordshire; West Midlands – Monmouthshire, Gloucester, Hereford, Worcester, Shropshire, Staffordshire, Cheshire; North – Yorkshire, Northumberland, Durham, Lancashire, Westmoreland, Cumberland.
Source: Paper Mill Directory, 1865, 1914.

craft in the North, South East, and Scotland, where over three quarters of all mills in the nation were now located. Even within these regions mills tended increasingly to be situated in a handful of counties. In the West Midlands 55 per cent of the region's mills in 1865 were found in Gloucestershire, Staffordshire, and Cheshire, and by 1914 this figure had grown to 85 per cent.[5] Likewise, in the North 76.9 per cent were located in Lancashire and Yorkshire in 1865 and 85.1 per cent in 1914. As for the South East, Buckinghamshire and Kent had 56.3 per cent in 1865; by the First World War 70 per cent of the South East's mills were located in these two counties. The Scottish industry was also overwhelmingly based around the large Edinburgh market.

Further evidence of the growing concentration of the industry in the 'established' centres of the trade is found in the fact that the five most prominent paper-producing counties, not only remained the same between 1865 and 1914, but also became more important. These counties ranked in order of importance were Lancashire, Kent, Yorkshire, Edinburgh, and Buckinghamshire. In 1865, 41.9 per cent of the nation's mills were found in these five counties alone, and in 1914 48.1. In this period the North also bypassed the South East as the leading papermaking region, primarily because of the growing importance of the industry in Lancashire. At a time when the number of mills

[5]   All statistics in this section come from *Paper Mill Directory of England, Scotland and Ireland* (1865ff).

in the industry was declining, the number found in Lancashire actually rose from thirty-eight to forty-seven. The only other major papermaking county to emulate this in this period was Kent (thirty-four to thirty-six). It should be noted, however, that as the distribution of mills through the country makes no allowance for the output of these mills, it cannot be taken as an accurate measure of regional importance in the trade. Indeed, given that the newer, larger mills of the second half of the century were mostly located in the North, South East, and Scotland, using the distribution of mills between regions to determine the locational pattern of the industry, if anything, tends to underestimate the importance in terms of national output of these key regions. This contention is supported by the industry's employment figures reported in the 1871 Census, which attributed to Lancashire, for example, about 19 per cent of the industry's workers, but only eleven per cent of the mills.[6]

The growing importance of Lancashire and other Scottish and Northern regions in the nineteenth century stemmed from several distinct advantages that these regions held. Proximity to major ports such as Liverpool, Hull, and Leith meant that these regions were well located to receive large quantities of the new imported raw materials, most importantly esparto and wood pulp. Indeed, it is not surprising to find that these regions led the way in the introduction of these new materials. The presence of neighbouring cotton-mills with ample supplies of cotton refuse had also long been used by the paper manufacturers of Lancashire as sources of cellulose. The region was also blessed with large quantities of pure and soft water as well as the ability to produce steam power and chemicals at reasonable prices. It also benefited from prior industrial and infrastructural development that had endowed the area with a trained labour force, machinists, and disused buildings and other sights suitable for conversion to paper manufacture, not to mention road, rail, and canal networks that linked the region to national and international markets. The region's numerous streams were another asset, not only because water was an important input into the production process, but also because they made for the easy disposal of effluent associated with the large-scale use of wood and esparto for papermaking. On top of these supply-side advantages must be added the growing and affluent cities and towns of the north of Britain that provided local manufacturers with rich markets on their doorsteps.[7]

---

[6] A. H. Shorter, *Papermaking in the British Isles: An Historical and Geographical Study* (Newton Abbot, Devon, 1971), p. 159.

[7] *Ibid.*, pp. 149–51, and M. Wray, *The British Paper Industry: A Study in Structural and Technological Change* (London, 1979), pp. 29–31.

The development of the industry in the North represented a fundamental transformation in the trade. Where before the presence of a rapidly flowing stream or river had been paramount to the selection of a mill's site, closeness to coal supplies, markets, and ports now appeared to reign.[8] Given the expansion in size of the average mill, this change in priorities only acted to intensify the concentration of the industry around London, Manchester, and Edinburgh. The only exception to this trend was in the hand-made branch of the trade which continued to be located almost entirely in the Maidstone area in the South East and in the South West.

Cost structure

By the second half of the nineteenth century survival in virtually all branches of the paper trade demanded machine production; a fact of life that necessarily made the industry much more capital intensive than it had been just fifty years earlier when hand production had reigned unopposed. The expenses of acquiring the necessary machinery, however, were formidable. In 1861 an average one-machine mill required a capital investment of some £5,000; by 1904 that figure had risen to around £30,000.[9] Investment on this scale inevitably made fixed costs a prominent feature of the industry's cost structure. This is clearly seen in the costs reported at the turn of the century by two established mills making similar grades of paper, but at different ends of the country – Dickinson's Croxley Mill near Watford and the Guard Bridge mill in Fife. Although we are not told how the costs are calculated, and whether they include an interest charge for the opportunity cost of capital, or how depreciation is imputed, they do at least give a fairly good impression of the composition of the typical costs faced by most papermakers. The breakdown of both mills' cost structures is almost identical with fixed costs, in both cases representing about 20 per cent of the mills' annual total running costs.

The importance of fixed costs and the slow rate of capital turnover provided strong incentive for producers making standardised products to lengthen production runs in an effort to spread the fixed costs of production over a larger output and to reduce average unit costs. There was thus an ever-present tendency in the industry for producers to offset high overheads and the fixed nature of the investment by striving for continuous production.[10] Maximising output not only reduced the

---

[8] P. W. Lewis, *A Numerical Approach to the Location of Industry* (Hull, 1969), ch. 4.
[9] Shorter, *Papermaking*, p. 147; Spicer, *Paper Trade*, p. 106.
[10] See J. G. Cummins, 'Concentration and mergers in the pulp and paper industries of the

Table 1.2 *Costs at Croxley Mill and Guard Bridge Paper Company, 1900/2 (in percentages)*

Costs	Croxley Mill 1902	Guard Bridge 1900/1
Materials	65	65
wages	13.1	12
fixed	21.9	23

Sources: Tariff Commission, TC3 1/84, p. 28; Weatherill, *One Hundred Years*, p. 119.

average fixed costs, but allowed manufacturers to minimise their losses in more dynamic ways. It encouraged them to show more interest in the economies to be achieved by involvement in supplying their own raw materials and disposing of their own products.[11]

A common, but rough, technique used to test for the presence of such economies of scale is Stigler's 'survivor test'. The underlying rationale for this test is that over time competition amongst firms of differing sizes tends to drive out those operating plants at relatively less efficient scales of production. By classifying firms in an industry according to their size and determining each of these classes' contribution to the industry's overall output at various junctures, the test allows one to identify the existence and consequences of possible scale economies in that industry. *Ceteris paribus*, if the market share of a given class of firm declines over time, then this strongly suggests that that plant size is relatively inefficient, primarily due to its inability to tap into scale economies. Conversely, those classes whose firms are able to operate at scales large enough to benefit from the presence of such economies will tend to see their share of the market increase rather than decrease with the passage of time.[12] Unfortunately, existing census and trade data for the British industry are not precise enough to attempt such a test. From 1904, however, the American *Census of Manufactures* does break down its national paper and board production

United States and Canada, 1895–1955', Ph.D. thesis, Johns Hopkins University (1960), p. 143; L. T. Stevenson, *The Background and Economics of American Papermaking* (New York, 1940), pp. 125–8; A. J. Cohen, 'The economic determination of technological change: a theoretical framework and a case study of the U.S. pulp and paper industry, 1915–1940', Ph.D. thesis, Stanford University (1982), p. 3/16; G. R. Armstrong, *An Economic Study of the New York Pulp and Paper Industry* (Syracuse, 1968), p. 68; and N. R. Lamoreaux, 'Industrial concentration and market behavior: the great merger movement in American industry', Ph.D. thesis, Johns Hopkins University (1979), p. 197.

[11] Lamoreaux, 'Industrial concentration', p. 217.

[12] G. J. Stigler, *The Organization of Industry* (Homewood, 1968), pp. 71–94. The test and its interpretation may be problematic where markets are not competitive and products not homogeneous. See F. M. Scherer, *Industrial Market Structure and Economic Performance* (Chicago, 1970), pp. 79–85.

14   Productivity and performance

Table 1.3 *Share of papermaking establishments in the US by value of product produced, 1904-1914*

Size	1904	1909	1914
Less than $5,000	3.0	2.6	1.4
$5,000–$20,000	7.8	7.3	6.1
$20,000–$100,000	33.4	26.1	22.1
$100,000–$1,000,000	51.9	57.5	60.9
$1,000,000 and above	3.9	6.4	9.5

Sources: Thirteenth Census of Manufactures (1905), table 11, p. 753; Fourteenth Census of Manufacturers (1914), table 11, p. 610.

between establishments of varying sizes, allowing us to conduct the 'survivor test' for the American paper industry. Even though only American data are used, this should give us some insight into the nature of the paper trade in general, as there is no reason to believe that findings based upon the American industry should differ markedly in principle from those experienced in Britain. The results of this test are presented in Table 1.3. There it is clear that between 1904 and 1914 establishments producing less than $100,000 worth of paper declined in importance from 44.2 per cent of the total to 29.6 in just one decade. The middle range group of establishments, between $20,000 and $100,000, fell most dramatically, while the smaller categories held up slightly better, mainly because many of these smaller establishments specialised in quality products for which the need for mass production was less pressing. Larger establishments also grew in importance. Million dollar establishments, for example, developed from producing a negligible share of overall value in 1904 to nearly 10 per cent in 1914. These changes are consistent with the notion that larger establishments were acquiring larger market shares because of the economies of scale they were capable of attaining. The speed with which these changes in market shares were taking place also suggests that the pursuit of ever-greater scale was a crucial ingredient to survival in the paper trade. Further confirmation of this is provided by the historical geographer, Peter Lewis, who has also claimed that such economies of scale were essential to survival in the British industry between 1860–5 and 1910–15. Indeed, using aggregate width of paper-machines as a proxy for scale of production, he has argued that the chance of survival was a linear function of size. For the period 1860–5 Lewis found that mills producing on average less than 278 tons per annum had less chance of long-term survival.[13] Some idea of the magnitude of the

[13] Lewis, *Numerical Approach*, p. 116.

Table 1.4 *Average size of paper-mills in the industry and in the newsprint and wax tissue branches of the trade in the US, 1895–1915 (000s pounds per day)*

Year	All branches	Newsprint	Wax tissue
1895	20.7	45.4	5.7
1905	64.9	169.8	16.0
1915	101.1	238.6	27.2

Source: Cummins, 'Concentration', pp. 38, 45, 55–7.

increases taking place in the size of average mills, once again in America, is given in Table 1.4. There it can be seen that average mill size grew between 1895 and 1915 by 388 per cent for the industry as a whole, 425 per cent in the newsprint branch, and even an amazing 377 per cent in the relatively small-scale wax tissue paper industry.

Fixed costs, though large, represented only a fifth of all costs accumulated in the paper-mill. The largest contributions made to overall production costs, however, were variable by nature. In Table 1.5 a breakdown of the cost structure of two different British firms in 1890/1, specifying the most important of the variable costs, is given. These show that the single most important item of both variable and total costs was the expenditure on raw materials. This, as with the chemical and wage bill, varied directly with the type and the quality of the raw material used in the mill, but even allowing for this degree of flexibility, the outlay made on raw materials was hardly likely ever to fall below a third of all of the costs incurred during production. This conclusion is confirmed by the cost data in Table 1.2 where the item 'materials', which included raw materials, chemicals, and coal, comprised, like the mills reported in Table 1.5, 65 per cent of total costs. Expenditure on materials and wages were also positively related to the grade of paper

Table 1.5 *Alexander Cowan and Sons Limited and Guard Bridge Company's costs of production, 1890/1 (in percentages)*

Costs	Cowans	Guard Bridge
Raw materials	42.8	42
Chemicals	15.8	17
Coal	7.5	7
Wages/Salaries	14.9	11
Freight	7.9	5
Others	11.1	18

Sources: Cowans Collection, GD1/575/8; Weatherill, *One Hundred Years*, p. 119.

being made, with the finer grades, employing only the best raw materials and more skilled labour, generally paying more for these inputs. Given the magnitude of raw material costs, the acute interest nineteenth-century paper manufacturers evinced in the search to find new and cheaper sources of cellulose, as well as the space devoted to the subject in later chapters, is understandable.

Market structure

Given the economies of scale that drove manufacturers to expand production to ever-greater levels as well as the slow but growing trend towards vertical integration in the industry, one might have reasonably expected a considerable degree of price-setting behaviour to have been exhibited by papermakers. Add to this the heterogeneity of paper, and the potential market power available to certain manufacturers would appear obvious. It should not be surprising, then, to find that students of the paper industry have indeed generally regarded the industry as being imperfectly competitive in structure. Despite this, however, many of these students have expressed discomfort at characterising the industry in such a way. Cohen, for example, remarked, obviously with some degree of amazement, that 'despite all these internal resource character-istics that fostered entry barriers and large-scale production, the industry was ... reasonably competitive'.[14] In her recent study of the American industry between 1900 and 1940, Ohanian has also argued on the basis of evidence derived from trade directories that the industry was very competitive with low barriers to entry. Indeed, the high rate of entry and exit in the industry between 1900 and 1940 is adduced as direct evidence that barriers to entry were not substantial despite the net decline in the number of firms and the growth in the average size.[15] Moreover, the exhaustive investigations of the Select Committee looking into the American paper industry of the early twentieth century also claimed to have found no evidence of either collusive or monopolistic pricing: a finding supported for the British industry in this period by Bartlett.[16] What fostered these competitive forces in the industry?

One of the most important factors compelling firms to operate with competitive pricing and profits, despite the advantages of scale, was the

[14] Cohen, 'Economic determination', p. 3/40.
[15] N. K. Ohanian, *The American Pulp and Paper Industry, 1900–1940: Mill Survival, Firm Structure and Industry Relocation* (Westport, 1993), pp. 51–67.
[16] Select Committee of the House of Representatives, *Pulp and Paper Investigation Hearings*, 5 vols. (Washington, 1909), vol. III, pp. 1971–88, and 3313–19 (hereafter *Pulp and Paper*); J. N. Bartlett, 'Alexander Pirie & Sons of Aberdeen and the expansion of the British paper industry, c.1860–1914', *Business History* 22 (1980), 19.

practice of grade-shifting. This practice involved the rapid switching of a mill from the manufacture of relatively unprofitable kinds of paper to those which appeared at the time more remunerative. The ability to do this is a distinct feature of the paper industry and its machinery, for with minor changes paper-machines can produce almost any grade of paper desired. Speeds can be altered and the concentration and composition of the stock varied and utilised in the same machine at little or no additional cost to the producer. The ability to grade-shift also stems from the fact that there is only a slight difference in the physical characteristics of different forms of paper. There is, thus, little difficulty involved in changing a mill designed for newsprint over to the manufacture of book paper. Even more dramatic shifts such as between cigarette and bible paper can also be made without encountering too many complications. As a consequence one machine can be used to manufacture a variety of paper products. Of course, the equipment necessary to make the usual product of the mill can in cases impose limits to this practice, especially when very high or low quality products are being made. For example, a cheap board-mill cannot easily shift to fine writing or some other luxury paper without first having acquired the appropriate equipment and experience. This fact endowed the manufacturers of such quality papers with considerable market power and a relative immunity to outside competition. Despite these limitations, in the vast majority of the branches of the trade where a standardised mass-produced product was manufactured, grade-shifting appears to have been a frequent occurrence.[17] Indeed, in 1889, J. A. Laurie, a super-intendent in a British paper-mill, saw the practice as an integral component of a paper firm's survival strategy: 'papermaking as a business is something that cannot be followed in any one line of practice. We must try one thing, and if that fails try another; in fact do anything, no matter what, in order to keep the machines going.'[18]

The most important effect of grade-shifting was to keep prices close to breakeven levels, simply because as prices increased (decreased) firms could easily enter (exit) the market and stabilise them. For this reason, Stevenson considered it 'one of the most important influences upon price behaviour in the paper industry ... It acts as a break upon runaway prices in a particular line and is a factor that all manufacturers are obliged to consider when establishing price policy.' He also believed it to

---

[17] Stevenson, *Background*, pp. 141–2; Cohen, 'Economic determination', pp. 3/45–7; and S. B. Karges, 'David Clark Everest and Marathon Paper Mills Company: a study of a Wisconsin entrepreneur, 1909–1931', Ph.D. thesis, University of Wisconsin (1968), p. 8.

[18] *Paper Makers' Circular* (hereafter *PMC*), 10 June 1889, p. 211.

be the best defence against the establishment of monopolies in the industry.[19]

The paper industry's market structure thus would seem to conform more to the description of the Bertrand model of imperfect competition, where firms charging prices above marginal cost risk losing all their business to other firms willing to undercut them, than to the Cournot, where each firm makes its output decision assuming other firms' behaviour is fixed. Although the Bertrand model generally presupposes a homogeneous product, it can also be extended to cover differentiated products, where the competition of similar though not identical products acts to eliminate the application of market power.[20] Our situation, however, does not exactly match that of Bertrand competition, since what operated in the paper trade to keep prices in check was not so much the competition between differentiated products, but the threat of cross-entry from a mass market of manufacturers with virtually identical production technologies. The theory of contestable markets, a natural extension of the Bertrand model to pre-entry behaviour, is of relevance here. A contestable market is one where established firms in an industry characterised by significant economies of scale and potential for monopolistic power are prevented from exercising this power by the threat of quick entry into the trade by rivals whenever excess profits are being made. This threat is made real by the absence of significant sunk costs and entry barriers.[21] This appears to match the situation in the late nineteenth-century paper industry. The only significant difference is that while entry into the trade may have been prohibitively expensive to those not already engaged in papermaking, it was a simple and relatively costless matter for other paper manufacturers, who chiefly specialised in another line, to produce temporarily another grade of paper where excess profits were currently being made. The threat of such grade-shifting forced paper manufacturers to keep their prices, as in the Bertrand model, close to competitive levels.

Demand

Beside costs and market structure, demand stands as a constant and important consideration in the industry's economic and technological development. Although its uses are manifold and diverse, paper is

[19] Stevenson, *Background*, pp. 150, 224; Ohanian, *American Pulp*, p. 67.
[20] J. Cubbin, 'Apparent collusion and conjectural variations in differentiated oligopoly', *International Journal of Industrial Organization* 1 (1983), 155–63.
[21] W. J. Baumol, J. C. Panzar, and R. D. Willig, *Contestable Markets and the Theory of Industry Structure* (London, 1982).

generally regarded as an intermediate good that derives its demand from that of the products for which it is an input. Some of the basic features of the domestic demand for paper between 1861 and 1913 can be seen in the estimated demand function:

log domestic demand = 1.24 − 0.595 log relative price of paper + 1.33 log income per capita

$$(1.70)(-6.50) \qquad\qquad (20.23)$$

SEE = 0.1202 R^2 = 97.3   F = 877.23   DW = 1.82

which is able to explain almost all of the variation in domestic demand in that period and which also indicates that the demand for paper is a price inelastic one.[22] The source of this price inelasticity is the derived nature of most of the industry's demand. Where paper is an essential input into a product, or the finished product itself has an inelastic demand, or alternatively, paper constitutes just a small portion of the total cost of producing the finished commodity, such an inelasticity of demand is only to be expected. These conditions appear to have held for most of the paper trade in the nineteenth century. For most types of paper, then, higher prices usually did not end up inhibiting demand so much as shifting it from one form or grade of paper to another. The newspaper proprietor's usual response to an increase in the price of one type of newsprint, thus, was not so much to reduce his overall consumption of paper as it was to turn to cheaper and lower grades of newsprint. In fact, this is what American newspapers did do during the Civil War when the

---

[22] *t* statistics are given in parentheses. The coefficient on the relative price of paper is significantly greater than − 1. The high R^2 might indicate autocorrelation, but the Durban-Watson statistic and an examination of residuals seems to rule this out. Likewise, Parks test, relating the square of the residuals to the explanatory variables, finds no proof of heteroskedasticity. There is also no suggestion of multicollinearity. Lack of data prevents the inclusion of paper's complements and substitutes in the regression. Presumably amongst the complements one would like to include the price of ink, writing implements, typewriters, postage, and all goods packaged in paper. Paper's substitutes are harder to identify. Prior to the advent of the computer, plastic, or other synthetic fibres, it is not clear what in fact could substitute, for example, for printing or writing paper. Older materials such as parchment, wood, cloth, or stone were hardly feasible. As for paper's packaging function, the closest substitute would have to be wooden boxes. Demand is calculated by subtracting exports from the output series in W. G. Hoffmann, *British Industry, 1700–1950* (Oxford, 1955), table 54. Because data on domestic paper prices are scarce, the relative price of paper in the above regression is calculated by dividing the average price of British paper exports by the average price of the principal industrial products (from which paper's various substitutes, whatever they may be, must have come). The years of 1870, 1889, and 1903 are outliers that have been omitted from the regression. The price data are given in W. Page (ed.), *Commerce and Industry*, 2 vols. (London, 1919), vol. I, p. 89; and B. R. Mitchell, *British Historical Statistics* (Cambridge, 1988), p. 772. Real national income comes from Mitchell and C. H. Feinstein, *National Income, Expenditure and Output of the United Kingdom 1855–1965* (Cambridge, 1972).

escalating price of the good quality printing paper, that they preferred, compelled them to switch to lower quality wood-pulp paper.[23]

The demand function estimated above also suggests that the demand for paper was significantly income elastic.[24] This factor played an important role in the development of the industry, as the demand for paper seems to have moved largely in tandem with the rising prosperity of Britain in the nineteenth century. By the second half of that century this steady growth of income per capita had started to filter through to all levels of British society, increasingly providing even the poorer of the industrial towns for the first time with sufficient incomes to purchase newspapers, books, and cards on a more regular basis. Indicative of this growing wealth was the appearance of the postcard, 76 million of which were already being delivered by the Post Office in 1872. By 1896 this figure has grown to 315 million; an increase that amounted to an average compound growth rate of just over 6 per cent per annum.[25] Such growth in demand meant that by 1890 total consumption of all grades of paper and board in Britain stood at 430 million pounds or about 18 per cent of all world consumption in that year. In terms of per capita consumption this meant that each inhabitant of Britain absorbed approximately 12.1 pounds of paper per year: the highest level of consumption in the world, larger than even the United States, whose average citizen only consumed 10.2 pounds per annum. In marked contrast, Europe's smallest user of paper in 1890 was Bosnia which had an annual consumption of less than 4 ounces of paper and board per head; or in other words, about 2 per cent of average British consumption.[26]

Perhaps the most visible manifestation of this increase in consumption was the proliferation and mounting circulation of newspapers. In 1841 541 different newspapers were published in Britain. Just under forty years later in 1880 this number had risen to 1,836. Similarly, newspaper circulation, which stood at 45.5 million in 1864, reached 174 million in 1896.[27] A good example of the new press that was responsible for a great

---

[23]   Students of the industry have all noted that the industry's demand is price inelastic. See, for example, Stevenson, *Background*, pp. 139–40; Cohen, 'Economic determination', p. 3/34; Guthrie, *Newsprint*, pp. 122–3.

[24]   The coefficient on the log of income per capita is significantly greater than +1. Hoffmann, *British Industry*, p. 93 also believes that the demand for paper is income elastic.

[25]   The Board of Trade, *Comparative Trade Statistics: Statistical Tables Showing the Progress of British Trade and Production, 1854–1895* (London, 1896), p. 28.

[26]   M. G. Mulhall, *Dictionary of Statistics*, 4th edn (London, 1899), p. 437. See also *Pulp and Paper*, vol. V, p. 3294.

[27]   *Post Office Directory of Stationers, Printers, Booksellers, Publishers, and Paper Makers of England, Scotland, Wales and the Principal Towns in Ireland* (London, 1872), p. 10; M. G. Mulhall, *The Progress of the World in Arts, Agriculture, Commerce, Manufactures,*

Table 1.6 *Paper consumption in Britain, 1882 (in percentages)*

Type	Percentage
Printing	53.5
Schools and offices	18.8
Account books	6.5
Letter paper	11.8
Sundry manufactures	9.4

Source: Mulhall, *Dictionary*, p. 436.

deal of this increase and which catered for the massive new newspaper-reading public that had begun emerging was the *Daily Telegraph*, whose circulation grew phenomenally from 27,000 in 1856 to over 300,000 by the late 1880s.[28]

Of course, the demand for paper was not restricted to newsprint. Assessing the relative importance of the demand for different types and grades of British paper is hampered by the absence of disaggregated consumption data. The earliest attempt to compile such data was made by Mulhall for the year of 1882. His estimates, however, excluded board. Moreover, given the absence of information on his methods and sources, one must be wary of taking his estimates too literally. Nonetheless, they do at least give an idea of the type of magnitudes involved. Mulhall's estimates, as percentages of overall consumption, are given in Table 1.6. From these figures it would appear that printing paper, which included newsprint and all book paper, assumed over half of all consumption in this period. Indeed, 90 per cent of all paper consumed in Britain in that year seems to have been bound for uses involving the conveyance and recording of the written word, whether in the exercise book, ledger, letter, or newspaper. Government consumption, however, was not a major source of demand for the industry. According to an 1857 parliamentary report, only 3,855,312 pounds of paper, or just 2 per cent of all paper produced domestically, were used in governmental, revenue, and parliamentary offices.[29]

Industries, Railways and Public Wealth (London, 1880), p. 91; M. G. Mulhall, *Industries and Wealth of Nations* (London, 1896), p. 82.

[28] In addition to the growth of income and the fall in the price of paper, several other factors such as changes in the law and the level of literacy in Britain played their part in the spread of the newspaper in the second half of the century. A. P. Wadsworth, 'Newspaper circulations, 1800–1954', paper presented to the Manchester Statistical Society, 9 March 1955, p. 20.

[29] This figure excludes inland revenue permits, excise labels and Post Office stamps. Returns of the Weight of Paper used in all the Government, Revenue and Parliamentary Offices, including printed and plain, during the Year 1857, *PP* XXXIV (1857–8), 333–4.

Table 1.7 *Structure of production, 1907*

Type	Percentage
Writing/Drawing/Envelope	13.6
Printing/Newsprint	52.4
Packing/Wrapping	21.7
Printed/Coated(not hangings)	4.4
Paste-/Card-/Mill-board	6.2
Other sorts	1.6

Source: *Census of Production* (1907), table 1, p. 624.

We are on more certain ground from 1907 when the first census of production appeared. Although the percentages given in Table 1.7 are for shares of production, not shares of domestically consumed production, this ought not to distort the picture too adversely. Unfortunately, the different categorisation of paper products in the trade statistics prevent us from modifying the figures given in the census by subtracting exports from them. However, as the share of production exported did not exceed 10 per cent, our inability to make such corrections should not pose too much of a problem. Despite this limitation, the census data confirm Mulhall's finding that around half of all paper was destined for the newspapers and printers. The census also gives information on wrapping paper and board which together appeared to take up about a quarter of all domestic production in 1907. Mulhall's estimates had not recorded wrapping and packing paper separately, but had listed it under sundry manufactures. Assuming that all of this category was in fact wrapping paper, this would make the maximum share attributable to this product in 1882 about nine per cent. Even allowing for a significant margin of error in Mulhall's figures, it does appear as if this branch of the trade grew in importance in the last twenty years of the century. This is not surprising as it is in this period that the wrapping of mass-produced products, especially foodstuffs, came of age.

Was the export market important to British manufacturers? Table 1.8 gives some idea of the breakdown of British paper exports in 1865, 1885, and 1905. It should be noted that the 1885 and 1905 data did not contain a separate category for brown and packing paper which for these years was listed as unenumerated. The most important component of British exports was writing/printing/envelope paper which represented at least two-thirds to three-quarters of all tonnage exported. This category's dominance strengthened over the latter half of the century so that by the first decade of the twentieth century, it was responsible for nearly four-fifths of all paper exports from this country. By contrast, Britain's

Table 1.8 *British paper and board exports, 1865, 1885, 1905*
(in percentages)

Type	1865	1885	1905
Writing, drawing and envelope paper	68.9	74.5	78.4
Brown and packing paper	18.7		
Paste-, mill- and card-board	2.9	4.0	5.1
Unenumerated	9.5	21.5	16.5

Source: *Annual Statement of the Trade and Navigation* (1865, 1885, 1905).

various boardmaking industries contributed only a small portion to the nation's exports, although a rise in their share is also noticeable. An implication of the rising shares of these two sectors is that Britain's brown and packing paper industry must have, in turn, been exporting a smaller share. The exact magnitude of this decline cannot be ascertained because of that industry's disappearance from the trade statistics after 1865. Nevertheless the fact that unenumerated papers in 1905 made up a smaller proportion of total exports in that year than did brown and packing paper in 1865 suggests that the share of brown and packing paper must have perforce fallen between 1865 and 1905.

Over this period the volume of British exports expanded rapidly. In 1865 total exports of paper and board excluding hangings totalled 141,075 cwt; by 1885 it was 733,110, and by 1905 it had passed the million mark to stand at 1,226,736 cwt. This represented a compound growth rate of 5.6 per cent per annum. Despite this growth in foreign demand for British paper, exports, however, remained a small part of the British industry's total output, varying between 7 and 10 per cent of that figure. The empire absorbed the greater part of these exports: 79.3 per cent in 1865, 79.1 per cent in 1885, and 68.4 per cent in 1905. The most important markets were the Australian colonies which together took up 49.2, 57.2, and 20.5 per cent of Britain's paper exports in 1865, 1885, and 1905 respectively. British India was another major market, usually being the final destination for between 15 and 20 per cent of Britain's paper exports. Although the importance of the imperial market-place to the British papermaker is unquestionable, the fact that these markets' share of the mother country's paper appears to have fallen as the century unfolded implies that if these were in fact 'soft markets' as some have argued, then the industry's reliance on them weakened rather than strengthened after 1865. Outside the empire the most important buyers were the United States in 1865 with 2.8 per cent and France in 1885 and 1905 with approximately 6.5 per cent of all exports in both years.

Summary

In this chapter some of the basic supply- and demand-side features of the nineteenth-century paper industry have been examined. It was found that by the late Victorian and Edwardian period most of the industry was clustered into three areas of the country: the South East, the North of England, and Scotland. Even within these regions paper-mills tended to be located predominantly in the counties of Lancashire, Kent, Yorkshire, Buckinghamshire, and in the environs of Edinburgh. As the century progressed the industry increasingly became more concentrated in these vicinities where easy access to ports, coal supplies, and large markets were assured.

Analysis of typical production costs revealed that the single most important component in the papermakers' cost structure was the expenditure on raw materials. Naturally enough then, much of their innovative efforts tended to focus on ways to reduce these outlays. The large investment needed in fixed assets together with the relatively low capital turnover, however, also had important consequences for survival in all but the highly specialised luxury end of the trade. Individual producers were under constant pressure to expand output and sales in an effort to reduce unit costs. The data available from the United States suggest clearly that larger firms generally survived better and became more important in the industry.

Despite the importance of scale, which might have given some firms in the industry monopolistic power, one finds instead that the late nineteenth-century industry was remarkably competitive, performing in many ways like a contestable market. The practice of grade-shifting, whereby paper-mills quickly shifted from the production of one type of paper to an entirely different one – a practice made possible by the similarity of the machinery in all mills as well as the chemical structure of all paper types – prevented monopolies from being established and compelled paper manufacturers to keep their prices at competitive levels.

The demand for paper was largely a derived one, depending greatly upon the demand for the final products in which it was a raw material. As a result it tended to be price inelastic. It was also income elastic. The rapidly growing demand experienced in the nineteenth century allowed producers to concentrate on reducing production costs and take demand and its generation very much for granted. Although paper found uses in a variety of places, the most important sector of the trade was that which was produced for printers and newspapers. This type of paper normally took up over half of all paper consumed in the kingdom.

British paper was also in demand overseas where the most important markets where the colonial ones. Although growing in magnitude over the century, production for export, however, never exceeded 10 per cent of all British output.

Thus far we have ignored one important determinant of the industry's structure and potential for further development: technology. In the next chapter we turn our attention to this factor.

2   Technological change

An important feature of any industry is its technology and the process whereby it changes. The aim of this chapter is to provide some understanding at both a theoretical and factual level of this important facet of our industry in the late Victorian and Edwardian era: a period in which British industry in general has been castigated for its failure to keep up with its main competitors technologically.

The first section of this chapter will give a brief survey of the rival theories of technological development, followed by an analysis of the structure and nature of technological change in the paper industry. The process of technological change in the industry in the second half of the nineteenth century can best be described as a gradual accumulation of technological knowledge rather than a process characterised by discontinuity.[1] In the third section the origins of technological progress in the industry are more closely investigated. In particular, the importance of innovation resulting from knowledge acquired in production is emphasised. Following this, a theoretical exploration of some of the factors which affect these origins is undertaken. The findings of this exploration are of great relevance to later chapters where the diverging technological performance of the American and British industries are examined. In the final section a profile of innovative activity in the trade is constructed from British patent data.

Technological development

At the heart of industrial decline and success lies technological change. Our understanding of this important process, however, is far from complete. As Jewkes, Sawers, and Stillerman noted in their seminal book, *The Sources of Invention*, 'there seems to be no subject in which traditional and uncritical stories, casual rumours, sweeping generalisations,

---

[1]   According to Coleman this cumulative advance of technology was also a feature of the industry's development in the first half of the century. *British Paper Industry*, pp. 192–3.

26

myths and conflicting records more widely abound'.[2] Nearly forty years on, Jewkes *et al.*'s observation retains its ring of truth. There are still amazing gaps in our knowledge of technological change that remain filled largely by platitudes. This certainly cannot be attributed to a lack of interest, as since the sixties the literature on technological change has greatly proliferated. In spite of this proliferation, very little is still known about technological creativity. Most attention on this topic in the literature has revolved around the relative importance of supply and demand for innovative activity. Two basic approaches are easily identifiable here: demand-pull and supply- (or technology-) push. Whilst in practice the difference between the two is far from clear cut, it is nevertheless a useful dichotomy for didactic purposes. This stems from the fact that both approaches offer a vastly differing emphasis to the role market signals play in shaping the nature of innovative activity.

The demand-pull approach fundamentally sees the innovator reacting to changes in the pattern of demand as expressed through the market. As such, it shows some consistency with the traditional assumptions of neoclassical economics. At its base is the belief that innovation is ultimately driven by needs and that these needs are expressed in the preferences of the consumer about what types of goods are desired (because they best meet these needs); those preferences are in turn reflected in individuals' demand functions. To start the innovative process off, the existing pattern of demand must be altered. Various factors can trigger such a change: rising incomes, income redistribution, changes in foreign demand and consumer tastes, and the demand for technological change created by technical interrelatedness being the most frequently adduced in the literature.[3] Interpreting this changing demand as a shift in the needs of the consumer, producers and innovators thus perceive an opportunity to augment private profit, as well as social utility, by supplying the appropriate good to cater for the new needs. From this demand for new commodities is derived an augmented demand for innovative activity to create the new products and processes. At its strongest, demand-pull theory goes one step further, asserting that advances in scientific knowledge are, in turn, induced by the demand for innovative activity that is generated by shifts in market demand. In short, this amounts to saying that the development of science is needs-driven and that the supply of innovation is highly elastic.[4]

[2] J. Jewkes, D. Sawers, and R. Stillerman, *The Sources of Invention* (London, 1969 [1958]), p. 33.
[3] *Vide* G. N. Von Tunzelmann, 'Technical progress during the Industrial Revolution', in R. Floud and D. N. McCloskey (eds.), *The Economic History of Britain since 1700*, 2 vols. (Cambridge, 1981), vol I, pp. 143–63 for discussion of some of these factors.
[4] Perhaps the best-known attempt to link innovative activity to demand-side forces is

Supply-push theories of innovation by contrast can best be distinguished from the demand variety by their downplaying of the influence of market factors. Rather than perceiving the process of innovation as essentially a reactive mechanism, supply-push theories see technological and scientific development occuring exogenously. This approach thus sees the amount of new technologies developed as a function of the amount and quality of inputs utilised. Amongst the usual inputs cited one typically finds the state of scientific knowledge, the availability of skilled labour and investible funds, the extent and nature of national education and technical training schemes, and expenditure on research and development. An improvement in any one of these factors is seen to lead to invention, so that for some an index of R&D expenditure has become a virtual proxy for the scale of technological change.

A different but very popular strain of supply-side (but not, strictly speaking, supply-push) explanation relates technological change to the scarcity of resources. Traceable back to Hicks, this view argues that technologies are chosen so as to economise on the utilisation of scarce, relatively expensive, factors of production.[5] However, Salter, following neoclassical theory, correctly complained that rational entrepreneurs under competitive conditions would always welcome any cost-reducing technology irrespective of its factor-saving bias. Any change in technique that did result from a movement in relative factor prices was thus no more than factor substitution.[6] While this is theoretically true, others have claimed that such arguments ignore reality. As Rosenberg noted: 'the notion of a wide range of alternatives readily available, as implied by drawing a smooth, continuous isoquant is largely a fiction'.[7] His argument is that if a firm has to commit resources to R&D to allow factor substitution, new knowledge is being created, and that this activity should be considered technological change.[8]

J. Schmookler's *Invention and Economic Growth* (Cambridge, MA, 1966). Using patent data from the nineteenth and twentieth centuries, Schmookler performed a time-series analysis of the railroad industry in the US in which he identified a strong lagged correlation between increased expenditure on capital goods and innovative activity in the railroad sector. The presence of a three-year lag, the time needed to bring new ideas to fruition as Schmookler claimed, strengthened the case that the innovator had been induced by the lure of expected profitability.

5   J. R. Hicks, *Theory of Wages* (London, 1932). Strictly speaking, this is actually an explanation from the demand-, rather than the supply-side of the market for innovation. However, as input prices are considered to be on the supply-side of the overall economy, it is not unusual to find them placed on the supply-side in discussions about technological change.

6   W. E. G. Salter, *Productivity and Technical Change* (Cambridge, 1960).

7   N. Rosenberg, *Perspectives on Technology* (Cambridge, 1976), p. 63.

8   Nelson and Winter also use satisficing behaviour and David path dependency to blur the factor substitution/technological change distinction. R. R. Nelson and S. Winter,

It would not be overly harsh to say that the debate between supply-and demand-side factors, whilst being the focus of much attention, has thus far borne no decisive result. Nor is it likely to. Both approaches seem to suffer from insurmountable limitations that mar attempts at greater explanatory power. These limitations stem mainly from the high degree of abstraction involved. Little insight is thus given in determining why, where, and when particular technological breakthroughs are made. Why has Firm A been able to innovate to meet new demands or exploit new opportunities when Firm B in the same industry and economy has not? Why was Firm A's solution the one to dominate when there were alternative paths available? Why did Firm A's response not occur earlier when the scientific breakthrough was first made and when the intensity of demand was equally, if not even more, acute? It is on these types of questions that demand-pull and supply-push theories shed little light. They are also the questions with which policy-makers are most concerned.

In recent years, an alternative approach to technological development, partially born out of dissatisfaction with the existing state of theory and consciously based on empirical findings, has attracted much attention in the literature. This approach, named technological accumulation by Pavitt, attempts as one of its main objectives to reconcile realistically the supply/demand-side controversy. The framework of this approach is best expressed in terms of Kuhnian scientific paradigms.[9] In addition to scientific paradigms Dosi suggests the existence of technological paradigms which he broadly defines as patterns or models 'for the solution of selected techno-economic problems based on highly selected principles derived from the natural sciences'.[10] Fairly evident examples of such paradigms are the technologies associated with the semiconductor, the internal-combustion engine, and oil-based synthetic chemistry. Each paradigm in addition is said to have its own set of heuristics, or principles for the identification and solving of problems, which define

An Evolutionary Theory of Economic Change (Cambridge, 1982) and P. A. David, *Technical Choice, Innovation, and Economic Growth* (Cambridge, 1975). Other useful works on induced innovation include W. Fellner, 'Two propositions in the theory of induced innovations', *Economic Journal* 71 (1961), 305–8; A. Ahmad, 'On the theory of induced innovation', *Economic Journal* 76 (1966), 344–57; and V. Ruttan and Y. Hayami, *Agricultural Development* (Baltimore, 1971).

9   G. Dosi, *Technical Change and Industrial Transformation* (London, 1984) and G. Dosi, 'Technological paradigms and technological trajectories: a suggested interpretation of the determinants and direction of technical change', *Research Policy* 11 (1982), 147–62.

10 G. Dosi, 'The nature of the innovative process', in G. Dosi, C. Freeman, R. R. Nelson, G. Silverberg and L. Soete (eds.), *Technical Change and Economic Theory* (London, 1988), p. 224.

various directions and manners of search compatible with the methodology and knowledge embodied in the reigning paradigm. In this schema a technological trajectory is defined as the activity of technological progress in one specific direction along the economic and technological trade-offs set out by the paradigm and its heuristics. Each paradigm offers a number of alternative trajectories or development paths. Progress along any one of these trajectories is characterised by a gradual accumulation in technological knowledge. The actual path, or trajectory, chosen is determined by a series of inducement mechanisms which include *inter alia* technological bottlenecks, a scarcity or abundance of critical inputs, changes in the composition and rate of growth of demand, changes in relative prices, and the presence or threat of industrial unrest. These factors then, broadly consistent with Rosenberg's 'focussing devices' and Sahal's 'guide-posts', influence both the rate and direction of technological progress, but only *within* the boundaries defined by the nature of the technological paradigm.[11]

In this way, the supply-push/demand-pull debate is resolved. Environmentally related factors such as demand and relative prices are seen to be crucial in shaping the rate of technological progress and the precise technology followed. They are also used as selection criteria at times of paradigmatic shifts. Conversely, supply-side features such as scientific knowledge determine both the opportunities of technological progress and the boundaries within which demand-side factors can operate. Moreover, the development of entirely new paradigms must perforce stem from fundamental advances in science.

Perhaps more important than its ability to place supply- and demand-side factors into proper perspective, the theory of technological accumulation also views the nature of technological development in a very different light from the neoclassical approach in which, it will be recalled, technology is generally applicable and easy to reproduce. In such a world firms choose innovations simply by drawing them from the current pool of technological knowledge taken as given. Technological accumulation, however, sees the process of technological choice and development in an entirely different way:

Instead we have firms producing commodities in ways that are differentiated technically from goods in other firms, and making innovations largely on the basis of in-house technology, but with some contribution from other firms, and from public knowledge. Under such circumstances, the search process of industrial firms to improve their technology is not likely to be one where they survey the whole stock of technological knowledge before making their technical

[11] Rosenberg, *Perspectives*, ch. 6; D. Sahal, 'Technological guide-posts and innovative avenues', *Research Policy* 14 (1985), 61–82.

choice. Given its highly differentiated nature, firms will instead seek to improve and to diversify their technology by searching in zones that enable them to use and to build upon their existing technological base. In other words, technological and technical changes in the firm are cumulative processes. What the firm can hope to do technologically in the future is heavily constrained by what it has been capable of doing in the past.[12]

Technological development is thus a localised, firm-specific, cumulative, and path-dependent process. As such, the theory of technological accumulation can also be related to the recent upsurge of interest in the application of evolutionary models to the study of technological change which in most cases also sees technological development in the same gradual, cumulative, and incremental light.[13] This type of gradual development is characteristic of a number of industries. One of these appears to be the paper industry. Studying the paper industry, therefore, may reveal valuable information not only about the development of the industry itself, but also in a more general sense about this important type of technological change. In the next section we focus in detail on the process of technological change in the trade in the late Victorian and Edwardian era.

Technological change in the paper industry

The latter half of the nineteenth century was a period in which new ideas about how to make better paper were continually being developed. The innovations in this period fell into two categories. Firstly, there were those that were associated with the search for, and development of, new raw materials for use in the production of paper. The rapid expansion of demand for paper over the nineteenth century severely strained the inelastic supply of the traditional material for papermaking, rag. As a response, an almost febrile search for new raw materials from the 1850s and 1860s began in earnest. This category of technological change will be looked at in chapter 4.

The other category of technological change in this period involved mechanical alterations and improvements in both the preparation of raw materials for production and the production of the paper itself. These changes contributed to the expansion of the volume of paper produced and the speed at which it could be manufactured. Moreover, it afforded

[12] K. Pavitt, 'International patterns of technological accumulation', in N. Hood and I. Vahlne (eds.), *Strategies in Global Competition* (London, 1988), p. 130.
[13] For example, G. Basalla, *The Evolution of Technology* (Cambridge, 1988); R. R. Nelson, *Understanding Technical Change as an Evolutionary Process* (Amsterdam, 1987); F. Rahmeyer, 'The evolutionary approach to innovation activity', *Journal of Institutional and Theoretical Economics* 145 (1989), 275–97.

papermakers greater control over the quality and texture of the final product. No longer were papermakers forced to make low grade paper simply because of the inferior quality of the machinery and raw materials available, or because of poor location. These factors combined to create an industry better able to meet the growing and variegated demand for paper of the late nineteenth and early twentieth century.

Understanding the nature of technological change in the paper industry calls for some familiarity with what paper is and how it is made. Chemically speaking, the basis of all paper and board is cellulose, a complex organic compound found in all plant tissue. When this cellulose is separated into individual filaments and then matted together, the sheets of cellulose it forms is in essence paper. Whilst all plant life is capable of yielding the necessary fibres to make paper, the actual number of materials from which a supply can be economically extracted is very limited. Traditionally the papermaker's favoured sources were linen and cotton rags, materials for which processing done prior to their arrival at the paper-mill had already separated the individual cellulose fibres from the impurities with which they are intertwined in their natural form. From the mid-nineteenth century on, other materials such as wood pulp, esparto grass, or straw were also used frequently, but these all required some additional processing by the papermaker. Once the raw material has been selected and acquired, the process turns to the actual making of paper. There are three distinct stages to this task, which in principle have remained fundamentally unaltered since the early days of hand and vat production. It is worth emphasising at this point that the type of raw material used does not change the basic procedures of papermaking.

The first of these stages involves the preparation of the raw material into a form (called stuff) suitable for transformation into paper. This is accomplished by beating and macerating the raw material into pulp. Although the traditional method of doing this had been by hammering or stamping the raw material in a mortar, by 1750 the Hollander beater, or engine, a machine which vigorously pounded the raw material against a set of sharpened metal bars, had largely supplanted the older technique. The purpose of this beating was not only to separate the strands of cellulose, the key ingredient of paper, from other unwanted substances also found in plant flesh such as the resinous and silicious materials that bind cellulose fibres together, but also to ensure that once liberated from these other substances these strands were shortened and fibrillated into the form most suitable for matting. In this preparatory stage pulp is also refined and treated by chemicals so as to produce a stuff with certain enduring characteristics such as whiteness and toughness. For example, bleaches and dyes are often added to the pulp at the time of beating so as to impart a particular colour to the resulting paper. After the beating is completed, the pulp is mixed with water to make a paper stock usually of a consistency of 99.5 per cent water and 0.5 per cent pulp.

Following the preparation of the pulp, the formation of sheets of paper begins with the so-called *wet end* of the process. Here the appropriately mixed stuff is taken from the vat or chest, where it has been stored, and allowed to run over wire-meshed cloth (with traditional hand-made paper, this was a wire mould), through which water drains by means of gravity, leaving behind a residue of intermeshed cellulose fibres which, although still wet and unfinished, is in structure paper. The next and final stage of production, the *dry end* of the process, sees the drying, finishing, and winding onto reels of the finished product. With traditional hand production this was an exacting and time-consuming process, requiring an army of skilled craftsmen working independently of the actual formation of the paper. With the advent of the paper-machine and mechanisation, however, the dry end of papermaking was for the first time united to the wet as a single continuous process.

Despite this transformation in the organisation of production, the basic technological principle underlying the paper-machine was no different from that of hand production. This can be seen in Figure 2.1 where a diagram of a standard nineteenth-century Fourdrinier paper machine is given. As it is drawn the machine runs from right to left. At its head is the stuff chest where the prepared stuff is stored immediately prior to use. A mechanical stirrer assures that whilst in the chest the stuff does not harden or deteriorate. When production begins a sluice on the chest is opened more or less according to the thickness of the paper desired, and the stuff flows from the chest into a revolving strainer where impurities such as knots, sand, and dust are separated from the pulp.

From there the strained pulp passes on to a leather apron, which directs it to an endless wire cloth conveyer belt, over which the web of paper takes shape. The wire is kept in motion upon a series of tiny copper rollers. These rollers, in turn, are held in place by a frame to which a slight but rapid lateral jerk is regularly administered by means of a crank; the shaking motion facilitates the release of the water and the meshing together of the individual pulp fibres. Most water in the stuff, however, is drawn out of the pulp by the force of gravity and the suction boxes placed under the far end of the wire. Even then, the drained-off water is usually not completely free of pulp, so it is collected beneath the wire in a large wooden save-all, from where it is returned to the head of the machine by mechanical means for re-use. The edges of the paper on

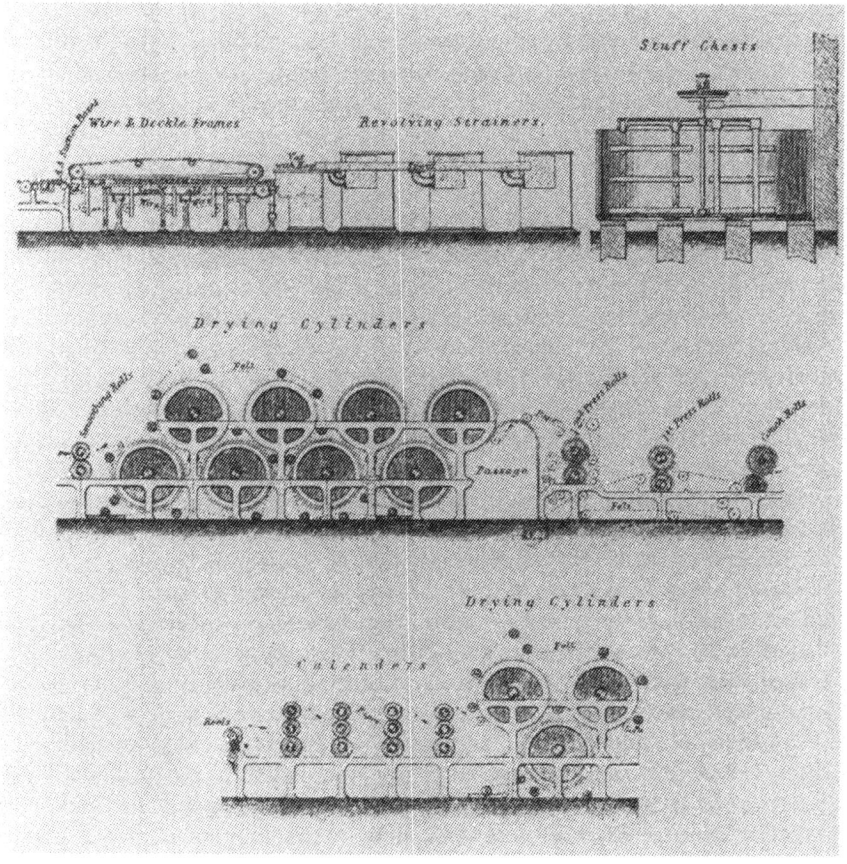

2.1   The Fourdrinier paper-machine

the wire are kept in place by belts or deckles of linen or rubber which stop the pulp from flowing off laterally before the fibres have set. At the end of the wire conveyer belt the paper passes through a set of couch rolls which transfer it from the wire cloth to a felt blanket moving at the same speed. In the meantime, the wire cloth goes under the press cylinders and returns to the start of the wet end where a new supply of pulp is obtained and the process repeated. From the wet felt the web of paper passes through a series of press rolls and drying cylinders, often steam-heated, which complete the drying process. Another series of smoothing rolls and calenders apply the finish to the paper which eventually is automatically wound onto reels ready for packaging and dispatch.[14] Without doubt the extension of the continuous production process from the wet end to the dry end that the paper-machine enabled was one of the major technological achievements of the industry in the nineteenth century. So successful and complete was this extension that the paper-machine, although actually an interrelated series of machines, is now usually described as a single machine.

This quality of interrelatedness in itself proved to be a significant source of technological change, because the technological imbalances created when improvements were made to one aspect of the machine tended to induce further changes to be made elsewhere in the machine in order that some sort of technological equilibrium be restored to the system. It also acted to concentrate attention on the existing technology and techniques, so that the weight of individual and collective innovative efforts of the firm and its employees tended to lead to firm-specific and incremental as opposed to discontinuous change in production methods. As Spicer noted in 1907 about technological change in the industry; 'there have been many alterations and improvements, but they have been for the most part continuations of the same idea. For seventy years the development has progressed by slow degrees at irregular intervals, along the path already marked out in 1830.'[15] Thus, despite continuous piecemeal improvement in the paper-machine in the latter half of the nineteenth century, the basic construction and workings of the machine remained true to, indeed in many aspects unaltered from, Nicholas Louis Robert's original machine of 1799.

Another feature of the technology which encouraged gradual accumulation rather than radical technological change was the paper-machine itself. The Fourdrinier paper-machine, the mainstay of modern paper-making, is a massive and expensive device. A 375 foot long 1,200 ton newsprint Fourdrinier, costing some $600,000 in the United States in 1921, for example, is reported to have required 60 freight cars for transportation.[16] Moreover, given the fact that each machine had originally been custom-built and once in place had been further developed to meet the specific needs of the firm, there was little possibility of a proper second-hand market in paper-machines being established. This is vividly illustrated by the dilemma faced by the Rocky Mountain Paper Company of Denver, Colorado, which, after it was

[14] For more detailed accounts of papermaking and the workings of the Fourdrinier paper-machine, see R. L. Hills, *Papermaking in Britain 1488–1988: A Short History* (London, 1988), ch. 2 and R. H. Clapperton, *The Paper-making Machine: Its Invention, Evolution and Development* (London, 1967), pp. 226–8.
[15] Spicer, *Paper Trade*, p. 66.
[16] A. J. Cohen, 'Technological change as historical process: the case of the US paper and pulp industry, 1915–1940', *Journal of Economic History* 44 (1984), 791.

forced to close down within a year of its opening in the first decade of this century, found that it had no other option than to sell most of its almost new paper machinery as junk and scrap metal.[17] Papermakers on the other side of the Atlantic likewise found it no easier to dispose of their unwanted machinery at a reasonable price. Faced by such limitations, firms intent on raising productivity and profits through technological change often had little other alternative than to improve the machinery they already had.

The objective of much of this incremental change in technology was to increase the speed and width of the paper-machine. This, however, was not just a matter of running existing machinery faster or welding on a few extra inches of steel, but required significant, if not major, innovations to be made. The problems encountered when increasing speed and width affected all aspects of the machine, even its basic construction. Faster and wider machines needed heavier and larger moving parts, and these, in turn, necessitated heavier gears; all of which only placed more stress on other parts of the machine. Other problems that needed to be confronted when augmenting speed were belt slippages, which became more frequent at faster speeds, and which caused an unwanted shaking of the machine as well as alignment problems and energy losses. Together these problems could result in greater downtime for repairs and damage to the paper web. Questions of the weight and balance of component parts of the machinery were likewise absolutely crucial. Table rolls, for example, had to be perfectly balanced or the resulting wobble could prove very disruptive to sheet formation. This balancing of the machine was complicated by the fact that wider and faster running machines usually required larger and heavier rolls. These larger parts also put greater stress on the drive mechanism.

The problems were not restricted to the structure of the machine. As the speed of a machine increased, the release of paper stock from the head had to be speeded up as well so as to maintain the consistency and thickness of the resulting paper. In fact, if the wire runs faster than the rate at which the stock is delivered, the stock drains so quickly that the cellulose fibres do not have the time to intertwine, producing as a consequence a weak and poor quality paper. Once on the machine, speed continued to cause problems, since there was less time for the water in the paper stock to drain through the wire. This was tackled by the development of various suction devices below the wire which drew off water faster than gravity alone did. By the First World War such

[17]   Hagley Museum and Library (hereafter HML): Thomas H. Savery Collection, Acc 1364 no. 5.

suction boxes, as they became known, had become standard features in all new Fourdriniers. These suction boxes, however, increased friction on the wire, and added to the machine's repair and energy needs. At the dry end of the machine faster speeds likewise meant that the paper web spent less time in contact with the dryers and calenders which therefore had less time to do their jobs. The easiest solution was to add more cylinders to the machine, but this only compounded the balancing and energy problems already mentioned. Larger machines, especially those with large drying sections, also created steam disposal and ventilation problems, which if not dealt with adequately, resulted in a rotten ceiling from which fragments of roofing and condensation could easily fall into the paper being formed below. Hoods to divert the stream and economisers to reclaim the heat and steam for use in heating the drying cylinders were the most common devices employed to circumvent these problems.[18]

Since the problems associated with increasing machine speeds and widths were manifold and interrelated, so that one problem could not be considered in isolation from the others, the ability of a firm or an industry to make such improvements to its machinery has thus come to be regarded by many in the trade as a fairly good proxy for technological progress. Not surprisingly, given the steady incremental nature of such technological progress, both of these proxies show a consistent rise over the second half of the nineteenth century. In 1876, Edward Lloyd Limited of Sittingbourne, Kent, at the time the largest paper manufacturer in the country, placed an order for a state-of-the-art machine that was an unprecedented 126 inches wide. By the end of the first decade of the following century, however, this achievement had long been surpassed with the width of the best British machines of the time standing at 150 inches and over. Similarly, where the normal speed of British machines was some 100 feet per minute in 1890, the average in 1907 was closer to 200 feet with the maximum in the country at 550 feet. Together such improvements in machine speed and width boosted the annual average rate of output per machine in Britain in the half century from 1861 to 1912 by well over 600 per cent, from 322 to 2063 tons.[19]

Evolutionary change also occurred on other fronts of the process such as with the development of better means of driving the machine and controlling the variation in its speed: an achievement largely brought

---

[18] Further discussion of the problems of increasing machine speeds is given in Cohen, 'Technological change', pp. 781–5, from which these paragraphs have been largely drawn.
[19] Spicer, *Paper Trade*, pp. 66–7, 143 and appendices VI and IX.

about by the abandonment of change wheels. Speed variation previously had been accomplished by positioning strips of felt daubed with resin on the two drivers so as to make one faster than the other. However, this procedure was dangerous and could communicate a shake to the machine that adversely affected the quality of the paper being produced. Engineers and attendants conscious of these problems worked on ways to eradicate them. By the turn of the century a long process of innovation had developed a driver that did away with all wheels and had a series of large taper pulleys, powered from a high-speed engine, which had been ordered in such a manner that each section of the paper-machine outside the cylinder group could be set to any tension desired. By this means speed could be altered by the machine attendants without difficulty or danger.[20]

Much of the technological change in the paper industry can thus be characterised by a gradual accumulation of technology. It was also generally a neutral process, affecting both capital and labour equally. As a result, improvements to the actual paper-machine rarely were labour-saving.[21] Variations in the amount of labour used also tended to be influenced only slightly by changes in the wage rate, since the manpower requirements of paper-machines were largely determined by the physical size of the machine actually used in the plant. A typical Fourdrinier, for example, was operated by between five and seven men depending on its width and length. Thus, as long as the machine was running, the size of the crew needed to operate the machine was fairly much fixed and inflexible to changes in the wage rates of its crew members.[22]

Sources of technological accumulation

There are four sources of innovation compatible with the type of technological accumulation experienced in the paper industry: the R&D performed by a special research department within the firm; R&D done by specialist R&D firms and institutions on commission; innovation carried out by capital machine producers; and innovation associated with the economies of practice (learning-by-doing, learning-by-using, learning-to-learn, and learning-about-payoffs). Of these four sources, the economics literature has focussed its attention almost exclusively on

---

[20] *Ibid.*, p. 67. For more details on this and other innovations, see Clapperton, *Paper-making Machine, passim.*

[21] Evidence for this can be found in L. P. Cain and D. G. Paterson, 'Biased technical change, scale and factor substitution in American industry, 1850–1919', *Journal of Economic History* 46 (1986), 153–64.

[22] I. Brotslaw, 'Trade unionism in the pulp and paper industry', Ph.D. thesis, University of Wisconsin (1964), p. 48.

the determinants of the optimal level of the first two and has made relatively little attempt to understand the latter two, especially the effects of learning. This is a pity, as it is precisely these latter two sources that figure most in the late Victorian and Edwardian era: a time when there were no independent firms anywhere specialising in the research and development of papermaking technology. Admittedly, some independent consultative chemists did exist, most notably the firm of A. D. Little and Co. formed in the United States in 1886, but their efforts were generally limited to testing the quality and strength of papers and were only infrequently called upon by papermakers to solve specific problems in production.[23] One problem was pollution from the mills. Chemists carried out experiments to find ways to reduce this pollution, which was always a source of great aggravation to those living and working in the environs of a paper-mill.[24] These independent chemists, however, had no systematic research and development programme and thus cannot be seen as a regular and reliable source of technological progress in the industry. Likewise, paper firms did not have any specific research departments or programmes. Any effort that originated from within the firm stemmed from the search by individuals for solutions to particular practical problems experienced in the mill. Such searches were generally undertaken after hours, almost as hobbies, and the solutions generated normally tested by incorporation into the everyday running of the plant, rather than in independent laboratories.

As with other industries, one of the most important ways innovation in paper-machine technology was transmitted through the industry was by the acquisition of new machinery. Paper-machine makers, therefore, played a crucial role, especially in the diffusion of best practice techniques through the trade. Machine makers, however, were rarely responsible for radical changes in paper machinery – not that there were many of these – but instead tended to develop and refine the ideas and techniques pioneered elsewhere, most commonly in the paper-mills themselves. They saw themselves as helpers to the papermaker, whether in the role of supplier of new technologies, or in seeing that the innovative papermaker (as well as they themselves) received a reasonable return for his commercially viable ideas. As the machine maker James Bertram and Sons of Edinburgh's catalogue issued in 1898 assured its readers:

---

[23] J. F. Magee, 'Arthur D. Little, Inc: at the moving frontier', paper presented to the Newcomen Society of the United States in 1986.

[24] Alexander Cowan and Sons, for example, hired the chemist C. J. Wahab for this purpose. Scottish Records Office (hereafter SRO): Alexander Cowan and Sons Ltd Collection, GD 311/7/31.

Any Paper-Maker desiring to develop new mechanical ideas may rely on their practical assistance; and the experience gained in their long connection with the paper trade, extending now over half a century, is freely placed at the disposal of their patrons. Every fresh invention relating to paper-making has their early and close attention, as will be seen from the accompanying pages where several novel machines are introduced, and minor improvements adopted to other machines, wherever their efficiency and value have been proved.[25]

Although there were certainly exceptions, this co-operation between papermaker and paper-machine maker appears to have been fairly pervasive by the late nineteenth century. The records of paper-machine makers and paper manufacturers clearly demonstrate the interrelationship.[26] As MacLeod has argued recently, this type of strategy, where papermakers shared ideas and profits (if any) with machine makers so as to capitalise on the paper-machine makers' technical and marketing expertise, constituted the most effective way of diffusing innovations through the industry. In this regard there is no reason to believe that British paper-machine makers and papermakers were failing each other in this period.[27] In fact, in the absence of such co-operation, the localised nature of innovation found in the industry at this time would have ensured that the diffusion of useful technology through the industry would have had to rely on the willingness of individual firms to share ideas, or on industrial espionage. These techniques, however, could only be successful in cases where the principle behind the innovation could be generalised enough to make it work in contexts other than that in which it had been orginally developed: a task that usually required the specialist talents of the paper-machine-maker.

In late Victorian Britain, paper-machine making was a small, highly specialised, industry. In fact, around the turn of the century there were only six major producers in the country.[28] Each of these operated in close contact with their customers. In addition to personal visits, this entailed the regular sending of advertisements, circulars, and catalogues to clients in order to keep them abreast of the new products offered by the firm.[29] Testimonials were also readily given and formed an important part of the marketing, as well as the diffusion, process. Typical of this was the Scottish papermaker, William Todd's, letter to

[25] SRO: James Bertram and Sons Ltd Collection, GD 284/25/1.
[26] For example, see William Todd of Springfield Mill's letter to George and William Bertram on 22 February 1883. In addition to paying an invoice for machinery received, Todd notes: 'there being also some ideas, one or two, on which we wished to speak with you'. SRO: Messrs James Brown and Co. Ltd Collection, GD1/575/8.
[27] C. MacLeod, 'Strategies for innovation: the diffusion of new technology in nineteenth-century British industry', *Economic History Review*, 2nd series, 45 (1992), 285–307.
[28] Tariff Commission, *The Engineering Industries* (London, 1909), vol. IV, paragraph 1252.
[29] For example, see Bertrams Collection, GD 284/25/1–11.

fellow papermaker, Lewis Evans of Dickinson, in February 1900 about the merits of a new pulping device called the Universal machine:

I see a Circular just received about 'Universal machines', a very good report from your firm undated. I shall be much obliged if you will let me know if you are still pleased with the worth of these machines and still prefer them to Kollergangs. I am told that you have also one of 'Corneth' breakers for broke: if so, do you like it as well or better than the 'Universal Machines'? Do you find the estimate of five to six HP for the Universals correct and do they or the Corneth clear out broke as well as Kollergangs, or if not so well, sufficiently so to put into beaters an hour or so before emptying without much of broken chips?[30]

Machine makers also arranged trips for customers to see their products successfully at work in other people's mills, and where further interest was shown were even prepared to lend some of their newer machines out for trial periods to paper manufacturers with no obligations or payments required until the satisfactory completion of the trial period. Alexander Annandale, himself a paper manufacturer and inventor of a patent box that he hoped to produce in conjunction with a paper-machine maker, contacted Todd in early 1903 about whether his firm would be interested in his invention. Todd was clearly interested in the device, but needed 'to see some tangible advantage to finally adopting it'. Todd proposed that he take the device for a trial period, since its suitability could 'only be decided by practical trial extending over some time, a month at the very least, perhaps more, and the final decision as its suitability for our work would have to lie with us . . . You may rely on the apparatus being given every chance and on everything possible being done to make it work successfully.'[31] Annandale readily agreed. But this was not just a ploy by Todd to postpone payment. In April the following year Todd once again wrote to Annandale, this time with bad news:

We have now had your patent box in for the three months trial arranged for and I am very sorry to have to report that I cannot see that it is any advantage to us. With thin paper it causes a good deal of worry and with ordinary substances we have been unable to discern any advantages either in closeness or strength.[32]

Soon afterwards, the machinery was removed and returned to its owner.
    The need for especially close rapport with the client stemmed from the fact that new machinery needed to be custom-built specifically for the needs of each individual mill, and so required assurances on the part of the machine builders that they understood the customers needs and were in a position to fulfil them at a reasonable price. Given the expense of paper machinery, it was only reasonable for the prospective customer to expect nothing less.

[30] Brown Collection, GD1/575/8, p. 823.   [31] *Ibid.*, p. 909.   [32] *Ibid.*, p. 924.

Whilst entire paper-machines represented an important part of a machine builders business, it was by no means all that they did. Most of their trade actually consisted of supplying parts either for repairs or to upgrade existing paper-machines. In fact, many of the smaller firms such as Edwin Amies of Kent did little else but this type of work. The same applied to larger firm like Bertrams which offered its clientele a wide-ranging catalogue of different machinery and parts. But even for Bertrams it was a relatively small-scale affair. Between 1860 and 1913, for example, the firm only managed to produce 118 complete paper-machines for sale – little over 2 a year. In good years such as 1876, 1890, and 1908 as many as 5 machines were made, though between 1 and 3 appears to have been the norm: hardly a scale on which to build a mass-production industry.[33]

Despite protective tariffs overseas, which stood as high as 45 per cent in America, British paper-machine makers seemed to have done fairly well on the export market – a fact which supports the claim that British capital producers were competitive.[34] Of the 118 machines made by Bertrams, 41, or about 29 per cent, were exported. The biggest markets were the Scandinavian, which took 21 of the machines, and the empire which took nine, although Bertrams' machines also ended up in Japan, Germany, Russia, and China.[35] In 1912 the company even erected an entire pulp plant, capable of producing 18 tons of dry unbleached pulp boards per day, in China for the Tonkin Pulp and Paper Mills Company. In addition to supplying the machinery and mill design, it sent an engineer for the purposes of superintending the erection of the pulp-mill machinery. He was under instruction to remain in China for as long as was necessary to get the plant running efficiently.[36] Less substantial work than entire plants or machines were also regularly and eagerly entertained; Edwin Amies, for example, supplying custom-made dandy rolls for paper manufacturers in France, Norway, Belgium, and Germany.[37] The point to be made here is that British paper-machine firms, like textile engineering firms, were very competitive, actively seeking new markets wherever possible, and in the process aiding the transfer of advanced papermaking technology to other countries and regions, notably Scandinavia, which lacked technological know-how.[38]

The other major sources of innovation were the insights and break-throughs that were achieved in the mill in the course of, or as a result of

[33] Bertrams Collection, GD 284/25/12.
[34] Tariff Commission, *Engineering*, para. 999.
[35] Bertrams Collection, GD 284/25/12.   [36] *Ibid.*, GD 284/25/11.
[37] Centre for Kentish Studies (hereafter CKS): Edwin Amies Records, Order book, U2852.
[38] K. Bruland, *British Technology and European Industrialization: The Norwegian Textile Industry in the Mid-nineteenth Century* (Cambridge, 1989).

problems encountered during, production. According to contemporary accounts this type of *ad hoc* tinkering and problem-solving represented a vital source of improvements, usually minor, that together boosted productivity within the industry. Given the intimate contact of many workers with the same continuous production technology for long hours, this was only to be expected. As a former employee of Guard Bridge mill in 1905 reminisced:

the various machine men had their own method of getting things to run right. It was usual to see bits of string and various odd weights attached here and there and the first half hour of the shift was usually occupied in getting these odd things adjusted. Some of them had special ways with the various pumps but all was to get a good run . . . each machine man having his own ideas.[39]

From such ideas originated many useful innovations. Likewise, engineer and machine shops located on the mill's ground, charged primarily with the task of repairing and keeping machinery running, found that to do so at times required innovations to be made. David Russell of R. Tullis and Co., for example, instructed his head engineer to work on the mill's boiler. In the process of this work, a Navy type boiler was developed that towered over the furnace with its tubes all sloping down so that they could get the advantages of the heat of the fire and save on energy. Unfortunately, the untimely death of the head engineer brought the project to a halt before its completion. The example, nevertheless, illustrates the importance of the machine shop for innovation. Most larger mills had such departments.[40]

To comprehend technological change in the paper industry of the late nineteenth century, it is clear that a better understanding of this source of innovation than the existing literature offers is needed. In the following section a start is made in this direction. Its objective is to explore the theoretical underpinnings of this source of innovation and suggest factors which might explain differences in intrasectoral and international rates of technological accumulation and learning.

Economies of practice and the benefits of learning

What do we know about the effects of experience on production? In a sense the intellectual heritage of modern economic thought on the

[39] L. Weatherill, *One Hundred Years of Papermaking: an Illustrated History of Guard Bridge Paper Company Ltd., 1873–1973* (Edinburgh, 1974), p. 33.
[40] C. Ketelbey, *Tullis Russell: The History of R. Tullis and Company and Tullis Russell and Co. Ltd., 1809–1959* (Markinch, 1967), p. 167; Weatherill, *One Hundred Years*, p. 19; J. Evans, *The Endless Web: John Dickinson & Co. Ltd., 1804–1954* (London, 1955), pp. 130, 137.

benefits of learning can be traced back to Adam Smith's analysis of the division of labour in which specialisation in the tasks of production, by permitting the continuous, full-time practice of a few simple operations, brought about improvements in the performance of those tasks. The first explicit reference to it, however, was made by Arrow in his famous paper of 1962 in which was first posited the notion of learning-by-doing. Learning-by-doing generally means the process whereby unit costs decrease as a direct result of accumulating experience. Learning-by-doing can only take place by performing an activity, the feedback from which gives rise to a response or stimulus by which the efficiency of performing the activity is increased. This may occur either by capability accumulation, which improves skills or organisation, or through pure technological accumulation which improves production equipment. Arrow's description of the phenomenon drew on the empirical work of Wright, who found that doubling the output of an airframe factory reduced average direct labour requirements by about 20 per cent, and Landberg's observation that the Horndal iron works in Sweden was able to raise productivity over a fifteen-year period at an average of 2 per cent per annum without new investment. Since there had been no significant change in production methods, Arrow concluded that in these cases 'steadily increasing performance. . . can only be imputed to learning from experience'.[41] In his model Arrow took cumulative gross investment as an index of experience because he regarded each new machine produced and put to use as capable of changing the environment in which production takes place. Yet Arrow also assumed that no learning takes place in the use of capital machinery once installed, and that all learning is a costless by-product of production and an entirely intra-firm process.

These are very strong assumptions which just do not allow for the feedback between machine makers and producers that has already been noted, for example, in the paper industry, or for that matter other forms of learning as well as intra-industry co-operation and collective invention: factors considered later in this chapter. The stringency of Arrow's assumptions probably stemmed from his desire to devise a model where technological change would be endogenous to firms.

Subsequent studies have adduced other interesting cases of learning-by-doing, such as Rapping's study of ship production during the Second World War, which supported the thrust of Arrow's work, as well as other cases which display other types of learning.[42] These include learning-by-

[41] K. Arrow, 'The economic implications of learning by doing', *Review of Economic Studies* 29 (1962), 156.
[42] L. Rapping, 'Learning and World War Two production functions', *Review of Economics and Statistics* 47 (1965), 81–6.

using in which experience with particular machinery leads to its
improvement; learning-by-learning in which previous experience with
innovation and other types of learning makes it easier to make further
changes; and learning-about-payoffs in which familiarity with a technology reduces the uncertainty about its benefits, and hence, the risk in
making further innovations.[43] Taken together, these learning effects,
called the economies of practice,[44] all share the characteristic of bringing
about technological change that is both local and neutral: local in that
they involve incremental and adaptive changes only, and neutral in that
they save equal proportions of all factors of production.

The benefits of learning, when known to the firm, carry other strong
implications for economic theory, including non-convexities, imperfect
competition, a multiplicity of equilibria, the suboptimality of myopic
maximising behaviour, and a potential justification for the infant
industry argument. History also becomes important because a firm's
ability to learn and develop depends critically on the technology in
place. One's initial choice of technology thus influences future outcomes. Making the right choice gives a country or firm a distinct 'first
mover' advantage, while a poor choice may lead them down a
technological dead end.[45] Moreover, in the presence of network
externalities, there is no way of predicting *a priori* if the chosen path is
the optimal one.[46]

A key implication of the economies of practice is that where they are a
factor a firm can augment its productivity and reduce costs if it can
increase its workforce's experience with the production process. Productive activity generates not only output in the physical sense, but also
information that is capable of improving production. In other words, the
more that is produced, the greater the benefits from learning. Whilst the

---

[43] N. Rosenberg, 'Learning by using', in N. Rosenberg, *Inside the Black Box: Technology
and Economics* (Cambridge, 1982); J. E. Stiglitz, 'Learning to learn, localized learning
and technological progress', in P. Dasgupta and P. Stoneman (eds.), *Economic Policy
and Technological Performance* (Cambridge, 1987); and for learning-about-payoff, see
R. Cowan, 'Nuclear power reactors: a study in technological lock-in', *Journal of
Economic History* 50 (1990), 541–68.

[44] The term can be traced to K. G. Persson, *Pre-industrial Growth: Social Organization and
Technological Progress in Europe* (Oxford, 1982), pp. 7–13.

[45] A. B. Atkinson and J. E. Stiglitz, 'A new view of technological change', *Economic
Journal* 79 (1969), 573–8; and Stiglitz, 'Learning', *passim.* David's 1975 classic explains
the overall rate and direction of technological development in nineteenth-century
Britain and America as the sum total of many such local and neutral changes. H.
Johnson, 'A new view of the infant industry argument', in I. McDougall and R. Snape
(eds.), *Studies in International Economics: Monash Conference Papers* (Amsterdam, 1970).

[46] W. B. Arthur, 'Competing technologies, increasing returns and lock-in by historical
events', *Economic Journal* 99 (1989), 116–31; P. A. David, 'Clio and the economics of
QWERTY', *American Economic Review* 75 (1985), 332–7.

direction of this relationship between innovation and production is undoubtedly correct, it tells us nothing about the expected magnitude of the outcome. Indeed, most models regard learning-by-doing as an automatic process that enables both efficient and inefficient firms to benefit equally from the experiences of production. This appears to be a weakness of much of the existing literature and represents a gap in our knowledge that simply weakens the effectiveness of the theory and its policy implications. For example, if two identical production processes are set up in different locations and ordered to produce exactly the same volume of the same good over the same time-frame, will the amount and nature of the gains from learning also be commensurate in both places? While there may be quite a few similarities stemming from the almost identical nature of the problems faced by both firms, one would also reasonably expect there to be noticeable differences. What are the sources of these differences? Apart from a brief allusion to cultural and mental attitudes by Stiglitz, the economics literature seems to be silent on this matter.[47]

As we have already seen, the benefits of learning emanate directly from participation in and familiarity with production. That is to say, it is to the worker, engineer, and manager actually engaged in the day to day running of production and the interaction between these agents that we must look for answers. In particular, the institutional setting in which they find themselves is especially crucial.

For didactic purposes it is useful to decompose the process of innovation from learning into five stages. The starting point is simply the act of production which creates valuable experience with the technology and the product for those engaged in production. On the basis of this experience, or learning, workers, familiar also with the problems and potentials of the process, begin over time to envisage solutions and directions in which the production process could be improved. But these ideas need to be acted upon and developed before implementation is possible. It is only with the implementation of these ideas that technological change can be said to have taken place. In brief then, the stream

Production → *Experience* → *Ideation* → *Activation* → *Innovation*

crudely characterises the process of innovation through learning. It also suggests the places where the stream can be diverted, blocked, or speeded up. There are three such points. Thus learning effects, for example, can be diminished only when (1) the nature and scale of production is insufficient to start the process off; (2) production does

[47]   Stiglitz, 'Learning', p. 143.

not lead to learning and to the attainment of experience; and (3) experience is not translated into innovation, because of either a failure to generate new ideas, or alternatively, an inability or unwillingness to act upon these ideas.

Although the economies of practice are intimately tied to all acts of production, certain types of production seem more amenable to the effects of learning than others. These are those forms of production which bring the worker into a more intense, routinised, and specialised contact with some aspect(s) of the process. The more people employed in applying a technology and the more times labourers work with that technology, the greater the probability of improvements being found. Continuous production line technologies with their high throughput and division of labour are especially susceptible to economies of practice. The point being made here is that where the level of demand permits it, anything that prevents firms from developing large production runs also lowers the benefits from learning.

The second juncture at which the process may be derailed or enhanced is with the transformation of production experience into learning. This obviously depends to a large extent upon the memory, perception, and inherent ability to learn of the individual workers involved. Such matters, while clearly important parts of the process, fall well beyond the scope of the present work and will not be pursued here except to note that just as experience in production increases one's productivity in producing, so experience in learning may increase one's productivity in learning. In other words, one can learn to learn, so that we could expect that one's ability to learn from production is enhanced over time.

Stiglitz has claimed that 'by assigning workers a variety of tasks, one may enhance the range of their learning capabilities ... they may then be more flexible in learning capabilities, better able to adapt to a wider variety of circumstances'.[48] This, however, would seem to run counter to the accepted view outlined above, that it is precisely that specialisation in, and resulting familiarity with, an activity that imparts the understanding necessary to transform it. This is not to say that Stiglitz's point is wrong. The benefits of broadening workers' experience is commonsensical enough, though there would appear to be a trade-off between broader and more specialised experience. Finding the optimal mixture of the two types of experience even at the best of times must be extremely difficult. Moreover, to assume the existence and search for this optimal balance of experience presupposes on the part of the firm an

[48] *Ibid.*, p. 131.

awareness of the economies of practice that may not have existed in late Victorian and Edwardian times. One further aspect which may affect the ability to learn is the rate of turnover of workers. Here the length of time that employees spend performing particular operations seems germane. If workers never stay in one place long enough to learn, then the experiences may be totally lost. Spells of unemployment and under-employment, casual and part-time employment, and frequent realloca-tion of labour to different tasks within the factory to meet short-term exigencies may undermine the acquisition of valuable experience.

Another way for a firm to expand its learning is to share its experiences with other firms in similar or related situations. We have already seen how capital producers and manufacturers in the paper industry often co-operated with each other to achieve beneficial results all round. Intra-industry co-operation also provided ample scope for similarly mutual gains to be had. According to Allen, many American iron and steel producers of the nineteenth century of their own volition shared their innovations with competitors in the belief that through such frank interchange of ideas and problems, the well-being of all concerned would be augmented. Indeed, in situations where status in the trade and competitive advantage can be had by demonstrating technological sophistication, or where trade and technological secrets are expensive to keep, or where release of the information will increase the value of assets and raw materials also held by the owner of the firm, or even where each individual in the trade has some cherished piece of knowledge which he hoards, there is strong incentive for this behaviour.[49] Such forms of collective invention, as Allen has called the phenomenon, play a crucial role in the gradual accumulation of technology, especially in industries such as the paper industry where the similarity of the technology used often led to the development of difficulties that were common to some extent to all firms. Such inter-firm contact is facilitated by trade associations, conventions, and journals. These organisations and pub-lications play an important role in creating a forum for discussion of shared problems and their resolution. As such, they represent a potentially valuable source of learning and expertise.

Equally important to the rate of innovation attained is the transition of what one has learned into working innovation. This may be disrupted at two stages: (1) with the generation of ideas, and (2) during the development and implementation of these ideas. A vast array of factors, many of them to do with individual psychology, influence the generation and implementation of technological ideas. This is one of the reasons

---

[49] R. C. Allen, 'Collective invention', *Journal of Economic Behaviour and Organization* 4 (1983), 16–21.

why technological creativity and change are such uncertain processes. Once again, however, we will restrict ourselves to the economic, technological, and institutional environment impinging on ideation and implementation.

In this regard the quality of the labour force is pertinent. To change the process one must first have a sound understanding of it. The task of the worker must be one that enables them to gain such an understanding. Where this task requires sophisticated knowledge beyond the capabilities of the worker, so that the task can only be performed by following a taught routine, or where the job is so isolated from the rest of the process that one cannot put the task into proper perspective, the opportunities for acquiring such an understanding may not be great. The level of education possessed by the average worker may in some ways enhance the extent of understanding attained, but with most production processes, especially in the last century, it would appear that formal scientific education beyond the rudimentary rarely pays off in terms of faster or improved learning capabilities. Practical experience picked up on the job is usually more relevant. Nonetheless, education may be still a worthwhile asset in that it may better the individual's ability not only to learn, but to understand and reason. This is particularly true of technical education which equips the worker with a good practical knowledge and understanding of his or her occupation. This type of worker is more likely to identify and resolve problems that occur in the day-to-day running of the firm. It would be safe to say, however, that the importance of education to learning effects probably varies from industry to industry, and within an industry from task to task.

Industrial relations and the institutional setting in which the employee finds himself working can also be germane. Working in a climate of industrial conflict characterised by confrontation between labour and capital and enmity for the owner, a worker may have no desire to help the boss out by improving his machinery; at least, not unless something is given in return. This raises an interesting question. It is a well-established aspect of virtually all theories of technological development that the ability to appropriate the returns to innovative activity is an important determinant of the level and rate of innovation that actually occurs.[50] This presumes that there is a clear proprietor of the innovation: *viz.*, someone or some firm that can claim to have fundamentally invented and developed the invention and can assert this property right.

[50] See for example, R. C. Levin, W. M. Cohen, and D. C. Mowery, 'R&D, appropriability, opportunity and market structure: new evidence on some Schumpeterian hypotheses', *American Economic Review* 75 (1985), 20–4, and Dosi, 'Technological paradigms', pp. 229–31.

In the case of innovation via learning it is not exactly clear who this 'owner' should be. In most instances it is the worker who actually sees the problem and finds its solution. The worker thus would seem to be the owner, yet the employer may claim that the development would not have been possible without the facilities and stimuli provided by the firm. Moreover, such activities, the employer would argue, are part of the worker's job for which he or she is already being remunerated. The problem then revolves around job control and description. On the assumption that all parties are cognisant of the effects of learning, we could expect it to become a bargaining chip in industrial disputes and negotiations. In particular it should affect negotiations over the level and determination of wages. The result of these negotiations will depend on the relative strengths and cohesion of unions and employer associations. It is worth noting, however, that unions, even those with an anti-technology bent, should not be too antagonistic to this form of technological change, as it displaces neither labour nor traditional skills. By contrast it is often a by-product of such skills. Faced with the prospects of further economies of practice, we could perhaps expect unions to agitate for piece-work wages, as the economies of practice allow greater output, and hence greater remuneration, for the same amount of effort. This is probably the best way short of productivity-tied wage rises for the workers to capture the fruits of their innovativeness and for managers to encourage even more of it.

Of course, some employers might resist such arrangements, preferring fixed hourly or daily wages, and where piece-rates exist a declining rate as output expands. Success for such employers is most likely where labour is disorganised, divided, or ignorant of the problem. In such circumstances the employer makes the benefits of learning part of the normal work routine and appropriates all the gains of the resulting productivity growth. When labour proves particularly troublesome, another alternative for the firm is to circumvent the problem and hire 'professionals' – engineers, scientists, technicians, etc. – specifically to improve both the product and its production. More skilled and better paid than the average worker, these new employees, more like junior management than workers, would be independent from trade unions and represent in embryonic form the beginnings of a separate R&D function within the firm. A confrontationist strategy, if it discourages workers from learning and innovating, may, however, prove counterproductive. In such circumstances if the benefits were considered important enough to the firm's well-being, employers might opt for remuneration plans favourable to its employees and the generation of further home-grown innovation.

Apart from piece-rates, other incentive-driven remuneration policies that could be employed would include profit-sharing plans, bonuses, and promotions for those who introduce or suggest innovative ideas.[51]

Managerial structure and the organisation of research and innovation within the firm and industry are also greatly influential in the generation and implementation of ideas. For all but the simplest of adjustments, improvements in production processes, especially where interrelatedness is a factor, will require changes to be made that disrupt production, possibly even bringing it temporarily to a standstill. Test runs may also be needed to retune the machinery.[52] All together the introduction of new ideas may impose a heavy cost aside from the costs of development in terms of lost output, so that any action upon a worker's idea would necessitate the approval of management. The internal structure and organisation of the firm, however, may make this difficult to obtain. In a sense this is an example of the more general need for channels through which information between the shopfloor and management can flow. One potential, and certainly not unknown, obstacle to such channels is the foreman who, perhaps out of jealousy or an overactive sense of self-preservation, deliberately discourages those under his charge from informing the management on how things could be run better, lest it reflects badly on him.

Decision-makers out of contact with their employees may be missing out on much potential innovation; a failure, which in terms of competitiveness, could turn out to be extremely costly indeed. It may sound strange that this would be ever allowed to happen, but it is conceivable when firms perhaps because of the perceived high set-up cost of internally generated research, ignorance of the economies of practice, or just complacency, decide to rely solely on the market and specialised R&D firms for their technology. The organisation and institutional structure of the firm can thus be extremely relevant to the realisation of the economies of practice.

---

[51] Support in the literature for such schemes are given in D. G. Blanchflower and A. J. Oswald, 'Profit sharing – can it work?', *Oxford Economic Papers* 39 (1987), 1–19; S. Estrin, P. Grout, and S. Wadhwani, 'Profit-sharing and employee share ownership: an assessment', *Economic Policy* 4 (1987), 13–52; and O. Wicken, 'Learning, inventions and innovations: productivity increase and new technology in an industrial firm', *Scandinavian Economic History Review* 33 (1985), 152.

[52] This was certainly the case in the paper industry. One American producer, for example, told a Congressional Hearing in 1909 that 'a few experiments, to be sure, have been tried with the sulphite process, but these have led to somewhat negative results, on account of the excessive costs of using a commercial plant for experimental purposes. This is one reason why the paper trade up to the present has not done more along this line.' *Pulp and Paper*, p. 1461.

A profile of innovative effort in the paper trade

Given the nature of technological change in the paper industry, what types of individuals got involved in innovative activity, and to what sort of projects did they direct their attention? One way of getting some answers to these questions is to use patent data. *Patents for Inventions. Abridgments of Specifications. Class 96: Paper and Paper-making* is an annual publication put out by the Patent Office which gives a comprehensive list of all patents concerning the manufacture of paper applied for each year. These lists also give brief descriptions of the workings of each patent applied for; a factor that enables one to assign each patent by type to specific parts of the production process or according to the nature of the claimed discovery. For the period 1855 to 1875 even more detailed information than this is available. Between 1855 and 1868 the *Chronological Index of Patents of Invention* supplied for each entry listed not only a brief description of the invention, but where furnished by the applicant, also the name, address, and occupation of the patentee(s). Between 1869 and 1875 this publication was replaced by the *Chronological and Descriptive Index of Patents Applied for and Patents Granted* which continued to provide the same information. Unfortunately, this publication and the information it supplied came to an end in 1875. Nonetheless, together both publications permit one to compile a basic profile of innovative activity in the British paper trade between 1855 and 1875.

It should be noted, however, that patents are at best imperfect measures of invention. There are a number of well-known limitations to their use for such purposes.[53] The most obvious of these is the simple observation that not all inventions are patented or even patentable. Indeed, many inventions for which patent rights are either difficult to enforce or obtain are simply not patented. For example, organisational changes, minor process innovations, and inventions that cannot be embodied in machinery or some other physical form – all changes which can obviously increase productivity – are often extremely difficult to capture in patents. The propensity to patent may also differ in different contexts. This is a particularly vexing problem for researchers who wish to study cross-sectional, cross-country, or very long-run patenting activity. The problem stems directly from the differing circumstances potential patentees find themselves in different countries, industries, or times. Institutional, technological, and cultural factors can play a big part in determining the proportion of inventions ultimately patented.

---

[53] K. Pavitt, 'Patent statistics as indicators of innovative activities: possibilities and problems', *Scientometrics* 7 (1985), 77–99.

Patenting is more likely, for example, in a nation or a time where the process of acquiring patents is relatively more straightforward and inexpensive and the rights of the patentee more strongly backed and supported by the courts. The nature of the technology also matters. Since patent rights are more easily protected when the patented invention involves capital equipment or physical products, these types of inventions tend to be more represented in patent counts than other types of inventive activity. Patent counts also do not tell us much about the value of the actual invention being patented, a major breakthrough receiving as much weight as a ridiculous dud. This complicates attempts to use patenting activity as a proxy for inventive output.

There are thus serious limitations to patent-based research. However, if one is aware of these problems and consciously attempts to minimise and contain their effects, patent data can still be a very revealing source of information. In our case the problem of differing propensities to patent is circumvented by the fact that our use of patent data is restricted to one industry, in one country, over a relatively short time-frame (fifty-eight years). In this period the most important alteration to the patent laws of Britain was the Patents and Designs Act of 1883 which besides introducing a new and much cheaper application procedure also modified the law to the advantage of the potential patentee in several particulars. For example, the Act of 1883 for the first time permitted patent applicants without fear of jeopardising their applications to exhibit their invention prior to filing an official patent application at international and industry exhibitions recognised by the Board of Trade. The Act also provided the patentee with the remedy of 'threat action', whereby a patentee, believing in good faith that an infringement of his patent has occurred, could threaten legal action against the alleged infringement without putting himself at risk of counter-litigation should the accused feel aggrieved by the patentee's threats or failure to commence and prosecute an action of infringement with due diligence. These changes and the simplification of the application procedure brought about a surge in the recorded number of patent applications being made in Britain.[54] These, however, were obviously not major concerns to those who took out papermaking patents; a class of patent which, as we will see later, did not experience the same type of surge in applications. The only other major change to patenting laws in Britain was the introduction of legislation in 1907 which required all patents applied for to be manufactured in Britain.[55] Though this affected foreign patenting in Britain, it had no effect on British patenting activity. There

[54] Mitchell, *Historical Statistics*, p. 439.
[55] Tariff Commission, *Engineering*, para. 1254.

Table 2.1 *Total and average number of patents for 10-year periods, 1855–1913*

Years	Total	Average
1855–64	472	47.2
1865–74	513	51.3
1875–84	463	46.3
1885–94	578	57.8
1895–1904	599	59.9
1905–13	566	62.9

Source: Patents for Inventions (1855–1913)

is thus little reason to believe that institutional changes or changes in the cost of acquiring a patent altered the propensity to patent enough to distort the information in the patent data regarding the temporal or structural pattern of patentable invention.

The 'value problem' is also reduced if we take changes in patenting as reflecting changes in the intensity of resources or effort devoted to innovative activities rather than a measure of the realised rate of innovation. This assumption is based on the belief that a fairly stable relationship exists between the use of resources in innovation and the desire to patent. After all, it is hard to believe that a firm or individual would commit resources to patenting activities in order not to innovate. Moreover, even if patents reflected only a small proportion of the total amount of innovative activity actually carried out, there is no reason to believe that their variation does not mirror changes in the resources devoted to all invention. In this section, therefore, changes in patent counts through time are taken as meaningful indicators of changes in the rate and direction of innovative effort. This is an interpretation that is commonly accepted and employed in the patent-using literature.[56]

Turning to the actual data, one finds that papermaking patent applications appear to grow over the entire period, albeit with a high degree of fluctuation, at an average compound rate of 0.58 per cent per annum. In Table 2.1 the annual average number of patents applied for over ten-year periods are given. It can be seen that with the exception of

[56] For example, see K. Pavitt, 'R&D, patenting and innovative activities: a statistical exploration', *Research Policy* 11 (1983), 33–51; R. J. Sullivan, 'The revolution of ideas: widespread patenting and invention during the English Industrial Revolution', *Journal of Economic History* 50 (1990), 349–62; K. L. Sokoloff and B. Z. Khan, 'The democratization of invention during early industrialization: evidence from the United States, 1790–1846', *Journal of Economic History* 50 (1990), 363–78; K. L. Sokoloff, 'Inventive activity in early industrial America: evidence from patent records, 1790–1846', *Journal of Economic History* 48 (1988), 813–50.

Table 2.2 *Share of papermaking patent applications out of all patent applications in Britain, 1855–1913*

Year	Percentage
1855	1.56
1860	1.85
1865	1.54
1870	1.64
1875	0.75
1880	1.05
1885	0.32
1890	0.35
1895	0.22
1900	0.25
1905	0.22
1910	0.15
1913	0.20

Sources: Mitchell, *Historical Statistics*, p. 439; *Patents for Inventions*.

the period 1875–84 the annual average number of patents rises over time. However, if compared to the increase in the number of all patent applications being made in Britain in this period, this rise can be seen to be a relatively slow one; patent applications in all industries grew at the much faster compound rate of about 4 per cent between 1855 and 1913. Papermaking patents, moreover, were not only growing at a slower rate, but also represented only a very small percentage of all the applications for patent protection filed in Britain at the time. As Table 2.2 shows, this share was at all times less than 2 per cent of the total, reaching a peak at 1.64 per cent in 1870. From then the figure fell steadily till papermaking patents represented a mere fifth of a per cent in 1913. There is no apparent reason for the failure of papermaking applications to keep up with the overall industry average. One implication of this comparison is that it appears that the enormous increase in overall patent applications that followed the implementation of the 1883 Act in 1884 was not reflected to the same extent by papermaking patents. Whilst the average number of all patent applications in Britain increased by over 200 per cent between 1875–84 and 1885–94, the number of papermaking applications filed only increased by about 25 per cent. For some unknown reason the 1883 Act does not seem to have had the same dramatic effect on the paper trade as on other industries. The importance of this curious finding is that for certain purposes we can still regard our patent data, despite the 1883 changes, as a continuous series. One such use is to see if the reported annual number of patent

Table 2.3 *Determinants of UK papermaking patents, 1860–1913*

	LALL (1)	LALL (2)	LRAW (3)	LRAW (4)	LRAW (5)	LRAW (6)
CONSTANT	1.53	3.00	5.19	6.37	4.65	4.63
	(3.47)*	(3.25)#	(6.97)*	(3.99)*	(14.36)*	(13.81)*
LCOST	0.261	0.149	0.576	0.486	0.666	0.651
	(2.63)*	(1.29)	(3.44)*	(2.43)#	(5.38)*	(4.86)*
LOUTPUT	0.243	0.097	−0.0601	−0.178		
	(5.47)*	(1.05)	(−0.80)	(−1.12)		
DUMMY		0.196		0.158		−0.0274
		(1.80)		(0.84)		(−0.31)
R^2	0.399	0.437	0.379	0.388	0.371	0.372
SEE	0.173	0.109	0.291	0.292	0.290	0.293
F	15.91	12.18	14.07	9.95	28.90	14.23
DW	1.75	1.76	1.95	1.93	1.94	1.95

Notes: LALL is the log of all patents, LRAW the log of all raw material patents. CONSTANT is the regression coefficient, LCOST the log of the average relative price of papermaking materials, LOUTPUT the log of industry output, while DUMMY represents the 1883 law change. t statistics in parentheses, where * = significant at 1 per cent level, # = significant at 5 per cent level. The Durban-Watson statistics for the regressions (1) and (5) indicate no autocorrelation, while regressions of the square of the residuals on the explanatory variables in both cases fail to turn up any statistically significant relationships, and hence proof of heteroskedasticity. An examination of the residuals and the square of the residuals likewise reveals no evidence of either autocorrelation or heteroskedasticity. Tests for multi-collinearity were also negative. Chow tests of equations (1) and (5) produced F values of 2.84 and 1.39, in both cases indicating at the 5 per cent level no structural break after 1884. The Franco-Prussian war and its immediate aftermath exerted a strong, though temporary, influence on the paper industry by effectively driving two of its main competitors, the French and German paper industries, from the international markets. To counteract the effects of this brief distortion, the years 1871 and 1872 have been removed from the regression.
Sources: Hoffmann, *British Industry*, T 54; *Patents for Inventions*; *Annual Statement of the Trade and Navigation*; Mitchell, *Historical Statistics*, pp. 722–4.

applications was linked in any way to conditions prevailing in the trade at the time. Although limited by the unavailability of industry data, the regression analysis of patenting activity presented in Table 2.3 does indeed suggest that they were. As regression (1) indicates, the level of overall patenting activity in the British paper trade does appear to have been positively related to industry output, suggesting a demand-induced or procyclical behaviour that others have noted for patenting activity in other contexts.[57] The number of patents applied for also appears to be linked to the average price of papermaking materials relative to the

---

[57] For example, Sokoloff and Khan, 'Democratization', pp. 366–7.

general price level in Britain. In fact, the approximate equality of elasticities of LOUTPUT and LCOST implies that demand- and supply-side factors had equal effects on patenting activity. This is a finding very much in contrast to most other studies which usually attempt to argue the dominance of one influence on innovative effort over the other. In reality, however, technological change is an amazingly diverse phenomenon, incorporating a wide range of new ideas, skills, and activities, and as such is unlikely ever to be adequately accounted for by monocausal explanations. In fact, taken together these factors explain only about 40 per cent of the variation in patenting activity in Britain; a figure, which even allowing for the incompleteness of the data, suggests that while they may be significant influences, they cannot be regarded as the sole ones on the rate of technological accumulation achieved. In terms of the strength of their influence, a 2 per cent increase in both the average relative raw material prices and demand would according to this regression bring about a 1 per cent increase in the number of papermaking patents taken out in Britain.

In equation (5) raw material-related patents alone were also found to be statistically linked to the relative price of raw materials, confirming the belief of contemporaries that the acute shortages of these materials perceived by the trade did in fact act as a fillip to the search for a new source of cellulose. An upward movement of 1 per cent in the price of papermaking materials relative to the general price level in Britain is seen to have induced about a two-thirds of a per cent increase in the number of patents to do with raw materials filed. Equation (5) explains around 37 per cent of all variation in this type of patent between 1860 and 1913. Equation (3) indicates, however, that this particular type of patent, unlike all papermaking patents taken together, was not significantly affected by the level of demand experienced. Demand influences were more pronounced on those patents concerning the actual paper-machine or the development of new products. Equations (2), (4), and (6) also appear to confirm that the 1883 change to the patenting law did not significantly influence papermaking patent applications in Britain. This is shown by the fact that in each of the regressions the DUMMY coefficient was not significantly different from zero, and that the Chow tests for equations (1) and (5) indicated no structural break after 1884.

The aggregate patent data can be broken down according to the part of the production process affected or by type. In Table 2.4 all British papermaking patents are classified according to whether their contents concern the preparatory, production, or finishing stage of the production process; while in Table 2.5 the same data are divided between those patents that involve a significant mechanical component and those that are non-mechanical in nature (i.e. chemical). Patents classified as 'Preparatory' include all new raw materials for making paper, new processes for using already known raw materials, new ways of treating the pulp during the preparatory process (e.g. bleaching, dying), as well as new ways of recovering chemicals spent in the process. It also includes all mechanical inventions and improvements affecting the transformation of raw material into paper stock from the dusters, washers, and rag cleaners of the rag room right through to the digesters and boilers of the boiling house and the beating and grinding equipment of the engine room. The classification 'Production' embraces all patents that are to do with the functioning of the paper-machine either at the wet or the dry end of the machine. As such, it includes *inter alia* innovations in wire, felts, deckles, alarm apparatus, pulp catches and savers, suction boxes, drying cylinders, calenders, cutters, rollers, and reeling and winding equipment. In this classification patents that concern the driving mechanisms of the paper-machine from prime movers to the transmission of motion through the machine have also been listed. In the third category, 'Finishing', one finds all improvements that add something to the basic paper, in the process making it a special and distinct product. This finish is usually added after the paper is made, but in some instances is applied at an earlier stage. This group of patents includes all finishing and sizing improvements as well as those special mixtures and chemical cocktails that when applied to paper give it a certain quality. This then would include patents such as those that suggest ways to make better security, marbled, or blotting paper. Entirely new paper products are also included in this category, but only as long as it is the papermaker in his mill who applies the necessary finishing, so that the new finish remains part of the process of paper manufacture. When it is a process carried out by someone in a paper-using industry, the patent has been excluded from the count.

The other distinction – between mechanical and non-mechanical patents – is fairly self-explanatory. The only problem is that in a number of instances a patent contained both mechanical and non-mechanical components. When this occurred, the patent was assigned to the category with which the most important aspect of the patent dealt. For example, a few raw materials patented also included a slight modification to existing pulping machinery that was needed to best extract the cellulose from the new raw material. In such cases, as it was the new raw material that was the essence of the patent and the mechanical component represented no more than a fairly minor modification of current machine technology, the patent was listed as being

Table 2.4 *UK patents according to stage of production process affected 1855–1913 (percentage in parentheses)*

Years	Preparatory stage	Production stage	Finishing stage	Total
1855–64	202 (59.2)	45 (13.2)	94 (27.6)	341
1865–74	244 (66.3)	65 (17.7)	59 (16.0)	368
1875–84	155 (48.7)	83 (26.1)	80 (25.2)	318
1885–94	246 (51.4)	127 (26.5)	106 (22.1)	479
1895–1904	192 (37.9)	155 (30.6)	160 (31.6)	507
1905–13	186 (36.7)	137 (27.0)	184 (36.3)	507
1855–1913	1225 (48.6)	612 (24.3)	683 (27.1)	2520

Source: Patents for Invention.

'non-mechanical'. Undoubtedly, some degree of error is to be expected here, though the number of patents involved are small and there is no reason to believe this error will be biased in any particular way.

Turning to Table 2.4, it can be seen that about half of all patent activity in British papermaking in the late Victorian and Edwardian period concerned the preparatory stage. The search for a commercially viable source of cellulose at this time largely accounts for this dominance. This percentage, however, fell fairly steadily in the last quarter of the nineteenth century, and markedly so from the mid-1890s as the use of cheap and effective means of producing paper stock from wood was popularised. The opposite appears to be the story for patents to do with paper-machines and the finishing stage. These grow both in terms of the number of such patents applied for, and their share of, all patenting carried out.

Looking at the nature of the patents, one finds an approximately even division between mechanical and non-mechanical over the period. As in Table 2.4, this hides the changes that were going on within the fifty-eight years surveyed. Reflecting the initial importance of the search for new raw materials to the industry and then its subsequent decline after substitutes for rag had been found, non-mechanical patents reach their peak share of all British papermaking patents between 1855 and 1864, only to have their prominence steadily lost over the following thirty years. The actual number of non-mechanical patents also fall at this time. A recovery of sorts in terms of both numbers and shares does take place from 1895, but this is mainly attributable to the proliferation of new finishes and paper products that appeared in those final decades before the First World War rather than the search for new papermaking materials. At the same time mechanical patents grew in number and

Table 2.5 *UK patents according to type, 1855–1913 (percentage in parentheses)*

Year	Mechanical	Non-mechanical
1855–64	128 (37.5)	213 (62.5)
1865–74	182 (49.4)	186 (50.6)
1875–84	176 (55.3)	142 (44.7)
1885–94	310 (64.7)	169 (35.3)
1895–1904	297 (58.6)	210 (41.4)
1905–13	250 (49.3)	257 (50.7)
1855–1913	1343 (53.3)	1177 (46.7)

Source: Patents for Invention.

share from the mid-century; a fact largely due to the ever-growing desire of papermakers of all descriptions for machines to run faster, more efficiently, and with fewer breakdowns.

Foreign patenting activity in Britain exhibited similar patterns. Approximately the same share of patents in both cases dealt with each of the stages of production. The preparatory stage as a locus of foreign patenting activity likewise also falls over the second half of the century. Similarly, foreign patent applications were evenly spread between mechanical and non-mechanical patents for the period as a whole. Mechanical patents, however, become more important as the period unfolds, except for after 1904 when they show a sharp decline. This decline appears to be related to the new patenting laws which required all patents to be produced in Britain. This change in the law, whilst reducing all foreign patenting, appears to have affected foreigners' propensity to patent mechanical innovations more noticeably than those for chemical or finishing processes. In any case, prior to 1905 foreign patents numbers and share of all patents applied for in Britain decline. Nonetheless, foreign patenting stood at about 20 per cent of all patenting activity in papermaking technology between 1855 and 1913 – not an insignificant proportion. A disproportionate share of all preparatory stage patents (22.9 per cent) and non-mechanical patents (23.6 per cent) were of foreign origin. To a large extent this reflected the fact that the search for new sources of raw materials for the industry was an international one with possible solutions emanating from all corners of the globe. Between 1855 and 1875 there were patent applications in Britain from sixteen foreign nations, although over 90 per cent of these came from only four countries. These were in order of importance France, America, Belgium, and Germany. France and America were by far the most important foreign patentees, together taking out about 80

Table 2.6 *Foreign papermaking patents in the UK by stage of production, 1855–1913*

Years	Preparatory stage	Production stage	Finishing stage	Total
1855–64	94 (31.8) [71.8]	14 (23.7) [10.7]	23 (19.7) [17.5]	131 (27.8)
1865–74	90 (26.4) [62.1]	17 (20.7) [11.7]	38 (39.2) [26.2]	145 (28.3)
1875–84	86 (35.7) [59.3]	32 (27.8) [22.1]	27 (25.2) [18.6]	145 (31.3)
1885–94	54 (18.0) [54.5]	20 (13.6) [20.2]	25 (19.1) [25.3]	99 (17.1)
1895–1904	25 (11.5) [30.5]	41 (20.9) [50.0]	16 (9.1) [19.5]	82 (13.9)
1905–1913	15 (7.5) [25.4]	12 (7.4) [20.3]	32 (14.8) [54.2]	59 (10.4)
1855–1913	364 (22.9) [55.1]	136 (18.2) [20.6]	161 (19.1) [24.4]	661 (20.8)

Notes: The figure for foreign patents as a percentage of all patents in Britain of that type is given in parentheses, and as a percentage of all foreign patents in square brackets.
Source: Patents for Invention.

Table 2.7 *Foreign papermaking patents by type, 1855–1913*

Years	Mechanical	Non-mechanical
1855–64	42 (24.7) [32.1]	89 (29.5) [67.9]
1865–74	56 (23.5) [38.6]	89 (32.4) [61.4]
1875–84	74 (29.6) [51.0]	71 (33.3) [49.0]
1885–94	52 (14.4) [52.5]	47 (21.8) [47.5]
1895–1904	53 (15.1) [64.6]	29 (12.1) [35.4]
1905–13	21 (7.7) [25.4]	38 (12.9) [64.4]
1855–1913	298 (18.2) [45.1]	363 (23.6) [54.9]

Notes: The figure for foreign patents as a percentage of all patents in Britain of that type is given in parentheses, and as a percentage of all foreign patents in square brackets.
Source: *Patents for Invention.*

per cent of all foreign patents. The empire in this period does not appear to have been a particularly fruitful source of patenting activity with only eight, or 2.7 per cent, of all non-British patents coming from it.

Between 1855 and 1875 information is available on the addresses and occupations of patentees. This material enables us to say something about the type of people involved in, and the geography of, patenting activity. To do this, it is necessary to divide the British industry into a series of geographical regions that meaningfully represent the distribution of papermaking across the country. There are eight such regions: (1) the South East (Hampshire, Sussex, Kent, Essex, Hertford, Buckinghamshire, Berkshire, Middlesex, Surrey, and London); (2) the South West (Cornwall, Devon, Somerset, Dorset, and Wiltshire); (3) East Anglia (Cambridgeshire, Norfolk, and Suffolk); (4) East Midlands (Derbyshire, Nottinghamshire, Lincolnshire, Warwickshire, Leicestershire, Rutland, Northamptonshire, Huntingdon, Oxfordshire, and Bedfordshire); (5) West Midlands (Monmouthshire, Gloucester, Hereford, Worcester, Shropshire, Staffordshire, and Chesire); (6) the North (Yorkshire, Northumberland, Durham, Lancashire, Westmoreland, and Cumberland); (7) Wales, and (8) Scotland. The distribution of patenting activity between these regions is given in Table 2.8. Where a patent was applied for by individuals from different regions, the patent was divided between each of the regions according to the percentage of patentees from that region. For example, if four individuals put their names to a single patent, and two of them gave addresses in the South East, one in Scotland, and one in East Anglia, 0.5, 0.25, and 0.25 would be added to the respective regional patent counts. In the absence of significant inside information about the contribution of the different individuals involved, this seems the only fair way of distributing the innovative effort between the different patentees. Table 2.8 also lists the average number of mills in each region as well as that number as a percentage of the nation's mills between 1855 and 1875. These are used to calculate the number of patent applications expected from each area of the country given the preponderance of paper-mills in the region. One could expect that patenting activity in the paper industry should reflect the geographical distribution of the trade. The figures given in Table 2.8 suggest that this is not the case. The chi-square statistic of 113.32 it generates allows us with 99 per cent confidence to reject the null hypothesis that the observed patent count did not differ significantly from the expected.[58] Looking closely at Table 2.8, it is clear that the South East's share of observed patents greatly exceeds its share of mills, while that of the North and of Scotland matches those expected. For the rest of the regions the amount of patenting activity experienced between 1855 and 1875 was much less than would have been expected from the distribution of mills in the country. The worst performer was the South West which with 8.4 per cent of the mills could only produce 2.2 per cent of the patents. What explains this pattern of innovative activity?

[58] For details of the chi-square test see M. Hamburg, *Basic Statistics* (London, 1985), ch. 9.

Table 2.8 *Observed and expected regional distribution of UK patents and mills, 1855–1875*

Region	Number of mills found in	National share of mills (%)	Observed number of patents	Expected number of patents
South East	94.5	26.8	179.7	99.2
South West	31	8.8	8.3	32.6
East Anglia	6	1.7	2.0	6.3
East Midlands	28.5	8.1	8.5	30.0
West Midlands	25	7.1	12.5	26.8
Wales	10	2.8	4.0	10.4
North	98.5	27.9	99.5	103.2
Scotland	59	16.7	55.5	61.8

Sources: *Chronological Index of Patents of Invention*; *Chronological and Descriptive Index of Patents*; *Paper Mill Directory (1855–75)*.

Table 2.9 *Observed and expected regional distribution of patents by paper manufacturers, 1855–1875*

Region	Observed number of patents	Expected number of patents
South East	44	39.5
South West	4	12.9
East Anglia	0	2.5
East Midlands	7	11.9
West Midlands	11	10.5
Wales	44.5	41.1
North	32.5	24.6
Scotland	4	4

Sources: As for Table 2.8.

In Table 2.9 the observed and expected distribution of patents applied for by paper manufacturers alone is given. These data generate a chi-square statistic of 14.01 which this time does not permit us to reject the null hypothesis. In other words, it cannot be said that the frequency of patent applications by papermakers of different regions significantly differed from the frequency one could have expected on the basis of the distribution of paper-mills between the regions. Scottish papermakers appeared to do the best and South Western the worst. Generally speaking, however, one can say that in regions where the paper trade was represented in significant numbers, the amount of innovative activity experienced between 1855 and 1875 was fairly much as could have been expected. Once again the South East performs well, although not to the same degree as in Table 2.8. This finding strongly implies that the key to the South East's dominance lay in its share of patents coming from non-papermakers. Indeed, 59.3 per cent of all patent applications originating from outside the trade were filed by South Easterners. The most probable explanation for this is that the close proximity of South Eastern mills to the enormous London market, rich with cash not to mention interested and talented people, provided a more fruitful setting for innovative activity than did other regimes where population and paper manufacturing were more sparse. In addition to these agglomeration economies, London, as the centre of international trade and finance, also possessed a large number of individuals who had detailed knowledge of other parts of the world and were thus ideally suited to make contributions to the search for new raw materials. London and the Home Counties thus emerge as very significant sources of innovative activity around the mid-century; a fact that lends support to the contention that by the Victorian era, the South East was already one of the most advanced and dynamic regions of the British economy.[59]

Another dimension of innovative activity can be gleaned from information on the occupation of patent applicants. About 60 per cent of all applicants reported their occupation. These are broken down into five categories: papermakers, mechanics, chemists, related trades, and unrelated trades. 'Papermakers' include the owners and all who find employment within a paper-, board-, card-, or wall-paper-mill; 'machinists' include all paper-machine makers as well as all other types of toolmakers or engineers (civil, consultative, etc.); 'chemists' include all trained as chemists or in one of the other practical sciences whether in an experimental, manufacturing, academic, or consultative capacity; 'related trades' is composed of newspaper men, publishers, stationers, rag merchants, size and wire merchants; and 'unrelated trades' embraces all not accounted for by the other classifications. In about 5 per cent of the patents more than one occupation was involved with the patent. These were counted in the manner that patents from more than one region were dealt with in Tables 2.8 and 2.9. It is worth noting in passing that about 75 per cent of these multi-occupational patents involved a papermaker working jointly with a machinist or chemist (40.9 and 31.8 per cent), while only 13.6 per cent had no papermaker listed. All patents had either a papermaker or machinist listed as a joint patentee.

The distribution of patents between occupation and type are given in Table 2.10. It is immediately apparent that patenting activity was diverse in origins. Papermakers make up the single most important group with

[59] C. H. Lee, 'Regional growth and structural change in Victorian Britain', *Economic History Review*, 2nd series, 34 (1981), 438–52.

Table 2.10 *Reported occupation of UK patent applicants and types of patent applied for, 1855–1875*

Occupation	Prep	Prod	Finish	Mech	Non-mech	Total
Papermaker	86	40	27	96	57	153
	[56.2]	[26.1]	[62.7]	[62.7]	[37.3]	
	(28.9)	(45.5)	(30.3)	(39.8)	(24.4)	(32.2)
Machinist	77	25	8	83	27	110
	[70.0]	[22.7]	[7.3]	[75.5]	[24.5]	
	(25.8)	(28.4)	(9.0)	(34.4)	(11.5)	(23.2)
Chemist	49	2	18	12	57	69
	[71.0]	[2.9]	[26.1]	[17.4]	[82.6]	
	(16.4)	(2.3)	(20.2)	(5.0)	(24.2)	(14.5)
Related trades	5	7	5	9	8	17
	[29.4]	[41.2]	[29.4]	[52.9]	[47.1]	
	(1.7)	(8.0)	(5.6)	(3.7)	(3.4)	(3.6)
Unrelated trades	81	14	31	41	85	126
	[64.3]	[11.1]	[24.6]	[32.5]	[67.5]	
	(27.2)	(15.9)	(34.8)	(17.0)	(36.3)	(26.5)
All occupations	298	88	89	241	234	475

Notes: These figures as percentages of all patents of those types are given in parentheses, and as percentages of each of the occupations' total patents in square brackets. 'Prep' is the number of patents concerning the preparatory stage, 'Prod' the production stage, and 'Finish' the finishing stage. 'Mech' is the number of patents mechanical by nature, while 'non-mech' are those of a non-mechanical nature.
Sources: As for Table 2.8.

half the patents concerning the preparatory stage and the rest being fairly evenly divided between the other stages. Over 60 per cent of patents were of a mechanical nature, with 45.5 per cent of all patents to do with the paper-machine being applied for by a practical papermaker. Machinists, who represent the third biggest group, naturally enough also focussed on mechanical innovations (75.5 per cent), their greatest efforts being devoted to the equipment in the beating room and the paper-machine. Together these took up 92.7 per cent of the patents. Chemists also focussed their attention on the preparatory stage, but tended to be engaged in mainly non-mechanical (chemical) innovations. Chemists were also disproportionately engaged in the finishing stage of production. In the related trades surprisingly little effort was devoted to papermaking innovations. This represented a fairly unimportant source of new ideas, making up only 3.6 per cent of all patents in this period. The efforts made by this group were evenly distributed between mechanical and non-mechanical patents as well as between the different stages of the production process. Most of their activity appears to have been directed towards the paper-machine, with such improvements comprising a not negligible 8 per cent of all patenting activity concerning the production stage. Finally, the second largest group applying for patents in the paper trade, 26.5 per cent of the total, was that which appears to have had no obvious link to the trade itself. Most of the efforts of this assorted collection of individuals focussed upon non-mechanical patents involving the preparatory stage. It was from such unrelated sources that many of the ideas – albeit mostly wild – for new raw materials for the industry originated. They were also significantly involved in suggesting new finishes for paper, but seem to have made relatively little contribution to the effort to improve paper machinery.

Although papermakers were actively engaged in all aspects of technological change from the mechanical to the search for new sources of cellulose, it is evident from the patent data that many other people of differing persuasions and backgrounds were participating in innovative effort that directly impinged on papermaking. This fact suggests that there was a fairly common awareness of the problems that the industry confronted. The influence of outsiders on the trade was most noticeable in the non-mechanical aspects of the preparatory stage of production. As already mentioned, the period 1855 to 1875 was one that was dominated by the search for a substitute to the industry's traditional raw material, rag. In finding a possible substitute, the papermaker did not have any special advantage. Rather, his advantage lay in the ability to perceive if an alternative was practically conceived and commercially viable. Most were not.

The single most important source of productivity growth in the industry, mechanical innovation, however, seems to have been the sole preserve of the papermaker, specialist machinist, and engineer. Of all paper-machine patents in that period, 81.9 per cent resulted from the activities of machinists, papermakers, and those in related mechanical trades such as wire manufacturing. This finding is understandable given that the detailed knowledge of the intricacies of the paper-machine that was needed to make improvements to it could only be acquired by direct contact and familiarity with the actual machine. This factor tended to render the resulting innovation cumulative and incremental by nature.

Summary

To understand the development of an industry requires one to also understand its technology. For this reason, this chapter has examined the process of technological change in the late Victorian and Edwardian paper industry. It found that technological change in this period fell into two major areas of activity: (1) the search for and development of new raw materials; and (2) piecemeal improvements to the paper-machine and other ancillary papermaking equipment. The first of these areas has not been considered much in this chapter, as it is the main focus of our attention in chapter 4. The other area of innovation, improvements to the paper-machine, represented as contemporaries were well aware a very vital and valuable source of productivity growth to the industry. Progress in this field was chiefly derived from extensions and improvements to the existing technology made possible by the learning and accumulation of technological know-how attained through the act of production. By nature, this was an evolutionary and continuous process. Such change was also largely incremental and firm-specific and tended to focus on increasing the speed and width of the paper-machine.

Changes to the paper-machine were generated and then transmitted throughout the industry by two mechanisms. The first of these was the paper-machine maker, who by absorbing all of the valuable improvements that individual paper manufacturers had made to their machinery in their mills and then adding their own innovations and refinements to them, enabled state-of-the-art technology to be embodied in the products sold to their customers. The other sources of change were the improvements that papermakers had learnt from the experience of production itself. Given the cumulative and incremental nature of much of the technological change actually experienced, this was obviously an important source of technological progress in the industry. The effects of learning on technological change, however, is not well understood in the theoretical literature. In particular, little attention in economics and economic history has been devoted to the determinants of the economies of practice. As a start, this chapter has offered a new way to approach this issue by exploring some of the factors that could influence the extent of innovation achieved through learning. It concludes that if one is to comprehend the rate of technological accumulation in an industry, one must look at factors that affect that industry's opportunity, ability, and willingness to learn from production. Such analysis would entail consideration of a wide selection of factors, ranging *inter alia* from institutional and organisational settings to technical education and industrial relations. In chapter 8 we will have reason to look at such factors again to see if they can provide us with some explanation for the growing technological divide between the American and British paper industries of the late nineteenth century.

Using patent data as a proxy for innovative effort, we have also been able to establish a basic profile of technological activity over the period of our study, and especially for the years 1855 to 1875. This source of data confirmed a growing secular trend in activity linked equally to the level of demand and price of papermaking materials. As the century progresses, there is also evidence of a change of emphasis in the patents applied for, with innovative attention seemingly shifting away from the search for raw materials and the preparatory stage to mechanical improvements to the paper-machine. A similar trend was observable in foreign patents applied for in Britain. Between 1855 and 1875 innovative effort was also disproportionately located in the main paper-producing areas of Britain: the North, South East, and Scotland. Of these the South East was the most important; a factor attributable to the agglomeration economies that London offered the region. In this period patenting activity was undertaken by persons of varied occupations. Of these different occupations, the papermaker was unsurprisingly the most important, especially when it came to improvements to the paper-machine, but others, including a great many who had no ostensible ties to the trade, were also actively involved in innovative effort. The contribution of such 'outsiders' tended to be almost exclusively devoted to non-mechanical aspects of the preparatory stage.

3   Performance

Over the course of the nineteenth century the British paper industry went from world leader to also ran. Yet it would be premature to conclude from this alone that the performance of the paper industry in Britain was somewhat lacking. Indeed, on the rare occasion that students of late Victorian and Edwardian Britain mention the industry, they speak of it rather favourably. For example, Ashworth described it as a 'rapidly expanding' industry[1] while Wray depicted the last quarter of the nineteenth century and the period leading up to the First World War as a 'period of expansion and prosperity for the paper industry'.[2] Similarly, Hoffmann, in his survey of British industry since 1700, chose to label it a 'non-typical' industry in which 'almost from the earliest years for which output statistics ... are available the smoothed rates of growth of output increase from year to year'.[3] According to his figures, from 1865 to 1913 the rate of growth of output fell below 4 per cent per annum on only two occasions (between 1876 and 1880 and between 1901 and 1905). These periods aside, the overall secular trend was one of rising rates of growth to 1900 followed by a gradual slowdown.[4] In spite of this impressive growth rate, at least by British standards, the British industry even in terms of output growth was still in the process of steadily losing its former international pre-eminence. So much so, in fact, that by the turn of the century Britain in terms of output had fallen to third place behind the United States and Germany with the gap between Britain and its two chief competitors ever widening. This gap between Britain and the United States is clearly depicted in Table 3.1.[5]

[1]   W. Ashworth, *An Economic History of England, 1870–1939* (London, 1960), p. 78.
[2]   Wray, *Study*, p. 34; see also Bartlett, 'Alexander Pirie & Sons', p. 30.
[3]   Hoffmann, *British Industry*, p. 183. Coleman also sees the growth pattern of the industry as 'untypical' (*viz.* its curve is non-logistic in shape). He attributes the paper industry's steady and continuously rising growth over the period to the development of new innovations and raw materials. D. C. Coleman, 'Industrial growth and industrial revolutions', *Economica* 89 (1956), 6–11.
[4]   Hoffmann, *British Industry*, table 54.
[5]   Spicer, *Paper Trade*, p. 230 and Hills, *Papermaking*, p. 187.

Table 3.1 *Output of paper and board in the UK and US, 1860–1914 (tons)*

Year	UK	US
1860/1	111,905	113,400
1870/1	169,023	344,600
1880/1	229,636	403,600
1890/1	411,483	834,800
1900/1	699,404	1,935,700
1905/7	899,900	3,765,200
1912/14	1,085,243	4,705,400

Sources: Hoffmann, *British Industry*, T54; Frickey, *Production*, p. 16.

Another sign of Britain's putative falling behind was the continuing trend for its paper imports to increase. From 1875 to 1895 imports rose from 41,000 tons to 543,000 and as a percentage of total domestic paper consumption from 27.7 to 52.9 per cent. Although this percentage fell in the following decades, the volume of imports entering the country after 1900 continued to be a major source of concern, and in 1907 still represented over 32 per cent of total consumption. From the late 1880s the balance of trade in paper (Table 3.2) also began to move strongly against Britain, despite the continued growth of exports. In the decade and a half before the First World War the magnitude of this deficit nearly doubled, so that by 1913 the value of imported paper and board exceeded that of exported paper to the tune of over £4 million.[6]

Yet it would be misleading to assume from these developments that a feeling of inadequacy or inferiority pervaded the industry. Indeed, as the official history of the trade association noted about the period, 'there is a note of almost jubilation in the Association's Annual Reports'.[7] Perhaps the most sanguine of the contemporary commentators, A. D. Spicer, an advocate of free trade, vigorously denied any failure on the part of the industry, and hence, need for protection:

in spite of the large increases in the importations of foreign papers and in spite of the decrease in the exportation of British paper, the material worked up in the British mills increased in ten years by sixty-four per cent. In face of this record there does not seem much reason for complaint on the score of foreign competition. It may be argued, however, that British paper-makers, although they are doing a larger business than ever before, may yet be doing it at too low a range of profit, or may, perhaps even be losing money. There may be cases

[6]   Wray, *Study*, pp. 34, 219–20.
[7]   A. Muir, *The British Paper and Board Makers' Association, 1872–1972: A Centenary History* (London, 1972), p. 26.

Table 3.2 *UK balance of trade in paper, 1861–1913 (in £000s)*

Year	£ 000's	Year	£ 000's
1861	189	1889	−305
1865	80	1893	−873
1869	167	1897	−1,957
1873	381	1901	−2,673
1877	303	1905	−3,316
1881	56	1909	−3,088
1885	14	1913	−4,010

Source: *Annual Statement of the Trade and Navigation* (1861–1913).

where this is the result, but on the other hand, there is as much proof that paper-makers are doing fairly well. The wills of private paper-makers, the dividends of paper-making companies, and the solid financial reputation of private concerns and limited companies engaged in paper-making, all offer abundant evidence that in spite of foreign competition the British paper industry, looked at from the manufacturer's standpoint, has not been altogether unremunerative.[8]

This may be so, but 'doing fairly well' and satisfaction with returns are terms that in themselves do not reveal very much about how successfully the industry was living up to its potential. In fact, they are words that are just as easily interpreted as reflecting the complacency that many commentators say enervated British industry in this period, as they are successful performance. In any case, assessing an industry's performance by its own past standards would seem a dangerous practice, not the least because it fails to allow for any fundamental changes in the economic environment that could have occurred in the interim.

Later students' cheery assessment of the industry may be equally unwarranted. Hoffmann's label, 'non-typical', reveals the problem with these favourable accounts. To see this more clearly, it is worth posing the question: in which ways was the industry's growth 'non-typical'? It was only with regard to the performance of other British industries at the time that the paper industry can be considered non-typical. As an observation, this may be quite true, but it seems inappropriate to assess one industry's performance by comparing it with an entirely different one, without at least first taking into consideration the differences and potential intrinsic to each. After all, some industries may exhibit rapid growth simply because they are starting from a low base, while others may be slow not because of inefficiency, but because they are mature,

[8] A. D. Spicer, 'The paper trade', in H. Cox (ed.), *British Industries under Free Trade* (London, 1904), pp. 204–5.

long-established industries which have exhausted both their technological possibilities and markets.

A more judicious measuring rod would be to evaluate the distance between foreign and British performance, as only this can give some indication of the extent to which the available opportunities have been seized by the industry. Output is also probably not the most appropriate variable to focus our attention on. The presumption that industries in a phase of fast growth are also making great strides in productivity is a fallacy that ignores the contribution of changes in total factor input. As Clapham reminded us back in 1938 about America's economic progress in general 'half a continent is likely in the course of time to raise more coal and make more steel than a small island, although this fact still surprised people between 1890 and 1910'.[9] This point applies equally as well to individual industries as to whole economies. As a consequence, no necessary correlation between output growth, export success, or the reduction of import penetration need be assumed. This may well have been the experience of the paper industry.

Labour productivity

To arrive at some worthwhile conclusion, therefore, the industry must be examined from the perspective of actual and potential development: *viz.*, to what extent were the opportunities available being exploited. One way to do this is to look at the comparative productivity of labour in different countries at different times. This is usually done by measuring physical output per head; or, in other words, the amount of labour needed per unit of output (in our case per ton of paper) in those countries. In industries where approximately homogeneous products are made in all producing countries, this method gives a crude comparison of efficiency. It is a crude method because, among other things, by focussing on the direct labour requirements needed to produce a unit quantity of output, it neglects quality factors which may reduce labour productivity. However, while this factor must be borne in mind, it would be erroneous to regard low physical output necessarily as a mark of high quality. Another problem with labour productivity is that it is quite an imprecise measure, reflecting the joint effect of a great number of influences on production. This is a problem shared by all measures of partial factor productivity. Calculations of labour productivity do not, for example, make allowances for changes in capital stock or energy use, so that rising output per head may mean that more machinery is being

---

[9]   J. H. Clapham, *An Economic History of Modern Britain* (Cambridge, 1938), vol. III, p. 122.

used rather than that an improvement in the productivity of the average labourer has taken place.[10] Labour productivity comparisons of the British and United States paper industries are given in Table 3.5. These figures have been calculated by dividing total output of all paper products produced in each country (though excluding the products of those industries that use paper, i.e. paper staining and box and bag-making) by the total number of individuals employed in the production of that output. As such, it is a very broad measure of labour productivity.

One problem in providing such estimates is the paucity of data. In particular, contemporaries left very little information about paper and board output in late nineteenth-century Britain. In most quantitative studies of British industry, therefore, the paper industry is lumped together with printing and allied trades. Students of the industry, however, have made attempts to trace the expansion of the industry from 1850. There are three such indices of paper production available: Spicer's, Hoffmann's, and Coleman's.[11] As already noted, the main problem confronted in attempting to estimate the level and growth of British output in the second half of the nineteenth century is the need to find a proxy that can substitute for the missing direct data. The official returns of the paper excise, on which all indices of paper production in the first half of the century are based, cannot be used as this source of data terminated with the repeal of the paper duties in 1861. From that date the three existing indices of paper production are constructed on the basis of the net imports of papermaking materials into Britain. This procedure is legitimised by the fact that the vast majority of the industry's raw materials in this period had to be imported into the country. Changes in the net imports of these materials are, therefore, taken to reflect changes in the output of paper. This assertion is corroborated by the observations of contemporaries, many of whom likewise shared the belief that increases in the import of raw materials directly resulted in commensurate changes in finished paper.[12] It is also confirmed by the existing data. Using information supplied in the 1907 and 1912 censuses of production, it can be shown that between these

---

[10] L. Rostas, *Comparative Productivity in British and American Industry* (Cambridge, 1948), chs. 1 and 2.
[11] Spicer, *Paper Trade*, appendix IX; Hoffmann, *British Industry*, table 54; Coleman, 'Industrial growth', 8.
[12] J. Kay, *Paper, its History* (London, 1893), pp. 40ff.; Mulhall, *Dictionary*, pp. 436–7; S. J. Duly, (ed.), *Timber and Timber Products, including Paper-making Materials* (London, 1924), p. 49; *Final Report of the Third Census of Production of the United Kingdom*, part 4, p. 286; Report of the Committee on Packing and Wrapping Paper, *PP* XV (1924–5), 637.

two censuses both the consumption of raw materials and the output of paper increased by approximately 20 per cent.[13] Hoffmann provides similar parallel developments for longer and later periods.[14]

Not surprisingly, given the similar basis of their construction, all three indices portray the same picture for the latter half of the century. This was a picture of steady and rapid expansion. Of the three estimates Hoffmann's appears to be the best, primarily because Spicers' series – and Coleman's which is identical to it until 1903 – contains violent fluctuations in output that are neither confirmed by contemporaries nor by the statistics of production of other industries. Moreover, Hoffmann's series has been corrected for an apparent overestimation of the amount of raw materials produced in Britain made by Spicer.[15]

For the United States the output figures used in our calculations come from Frickey's series for the industry between 1860 and 1914. These were based on material furnished by the Division of Forest Economy of the United States Department of Agriculture. The series represents the aggregate production of newsprint, book, board, wrapping, fine, and all other paper; and they were compiled or estimated by the Forest Service from census and other data.[16] The only noticeable difference between the data is that, unlike the British, the American includes paper hangings, but, as this sector represented only a very small proportion of the paper trade, the effect of this difference on the measurement of American productivity is likely to be minor. Information on German output and employment comes from Hoffmann's index of the paper-producing and finishing industries. The inclusion of the paper-finishing industry in the index does not cause any trouble for estimates of the growth rate of the German papermaking industry, because Hoffmann actually had no data for paper-finishing and assumed that it grew at the same rate as the paper industry.[17]

Data for labour input in America and Britain come from their respective censuses. The figures used measure all wage-earners and salaried persons working in the industry. One problem is that after 1900 US censuses stopped listing employment in the paper and the pulp industries separately. However, in 1880 employment in the pulp industry was less than 5 per cent of that of the paper industry.[18] Although this percentage would have certainly grown over the following decades, the inclusion of pulp-mill workers should not affect our labour

[13] *Final Report of the Third Census of Production*, p. 282.
[14] Hoffmann, *British Industry*, p. 279.   [15] *Ibid.*, p. 280.
[16] E. Frickey, *Production in the United States, 1860–1914* (Cambridge, MA, 1947), p. 16.
[17] W. G. Hoffmann, *Das Wachstum der deutschen Wirtschaft seit der Mitte des 19. Jahrhunderts* (Berlin, 1965), p. 375.
[18] *Tenth Census of Manufactures* (1880), p. 12.

Table 3.3 *Annual hours worked per person in the UK, US, and Germany,*
1870–1913

Year	UK	US	Germany
1870	2,984	2,964	2,941
1890	2,807	2,789	2,765
1913	2,624	2,605	2,584

Source: Maddison, *Dynamic Forces*, Table C9.

productivity calculations excessively. In any case this bias after 1900 strengthens some of the arguments raised later on, as it would lead to an underestimation of American productivity.

In addition to the number of workers employed, the amount of labour used at any time depends also on the number of hours actually worked. Information on hours worked in nineteenth-century industry, however, is sparse and notoriously unreliable. Maddison's figures (Table 3.3) for the whole British, American, and German economies between 1870 and 1913 suggest that average annual hours were similar and moved in parallel ways in these countries. These broad conclusions are supported by evidence from the paper industry. By multiplying the average number of days worked per year in both countries by the weighted average of hours worked per day (where the weights represent the share of the workforce reported to be working a certain number of hours per day), we can get some idea of the annual hours worked per year. The results are given in Table 3.4 Even allowing for error, our estimates of hours worked in the American and British industry throughout the entire period of the study are very similar. Since the hours of work in the American and British industries do not seem to have been markedly different, this factor cannot substantially alter our results, and thus has been excluded from our productivity calculations. It is also noteworthy that at all times longer hours were being worked in the paper industry than were being worked on average in both Britain and America. This is not surprising given the fact that there was strong incentive for papermakers to run their mills around the clock. To do so, workers in both countries were frequently called upon to work the 'long drag': a practice that could see them work up to 36 hours at a stretch and 144 hours in a week.[19]

An interesting aspect of the productivity histories of the British and American industries is the persistent gap in the level of labour productivity attained. Table 3.5 shows the scale of this disparity, as well

[19] Royal Commission on Labour – Textiles, Clothing, Chemical, Buildings and Miscellaneous Trades (Group C), *PP* XXXIV (1893–4) (hereafter *RC Labour*), 508.

Table 3.4 *Average annual hours worked per person in the paper industry in the UK and US, 1860–1907*

Year	UK	US
1860	3,350	3,336
1870	3,371	3,371
1880	3,340	3,384
1891	3,202	3,228
1901	3,212	3,179
1905/7	2,873	2,856

Sources: RC Labour, pp. 507–11; *Hours and Earnings*, p. 310; *Eighth to Fifteenth Census of Manufactures* (1860–1914); *First Report of the Commissioner of Labor*, pp. 388–91; UK censuses (1860–1912).

as displaying the relative levels of the German and British industries. From these comparisons it would seem that, allowing for some fluctuations due to the effects of each nation being at different stages of the trade cycle, the American industry consistently was about twice as productive as the British industry, and that German productivity levels, with the noted exception of the 1890 figure, were likewise persistently slightly below the British. This long-lasting Anglo-American productivity gap, which recent literature has also observed for the entire manufacturing sector, is something of an enigma; its dénouement may greatly enhance our understanding of both British and American economic development in the latter half of the nineteenth century.[20]

Turning to the average growth rate of labour productivity, one sees another dimension of Britain's performance in the late Victorian era. Over the whole period, and especially between 1880 and 1912, these growth rates in Britain appear comparable to those attained in the United States. The only period in which Britain seems to have been seriously lagging behind the United States was in the first decade and a half of the comparison, but as we will see in chapter 7, the diverging growth rates at this time were related more to choice of raw material and the differing labour requirements of each nation's mills than the poor performance of British papermakers.[21] In fact, in the final two decades

[20] S. N. Broadberry, 'Manufacturing and the convergence hypothesis: what the long run data show', paper presented at the ESRC Quantitative Economic History Conference at St Antony's College, Oxford, September 1992.

[21] This was possible because each papermaking material requires significantly different amounts of labour to be converted into a pulp suitable for papermaking. The disparity between American and British practice in this regard seems to have reached its peak around 1870 when, as Table 3.5 suggests, the labour productivity gap between the two countries stood at approximately the two-and-a-half to one mark. For further discussion of this issue, see chapter 7.

Table 3.5 *Comparative UK/Germany and US/UK levels of labour produc-tivity, 1860–1912*

Year	UK/Germany	US/UK
1860/1		169.0
1870/71		269.8
1880/81	117.7	195.0
1890/91	183.8	192.8
1900/1	139.8	181.6
1907	134.1	192.2
1912	117.4	199.7

Notes: Benchmark for levels from 1912 census and Chapman, *Work and Wages*, p. 237. Missing data in Germany are filled by interpolations between the two closest years. *Sources:* Hoffmann, *British Industry*, T54; Frickey, *Production*, p. 16; Hoffmann, *Wachstum*, T15–16; UK and US censuses (1860–1914).

of the last century, Britain, at least in terms of labour productivity growth rates, seems to have performed well *vis-à-vis* the United States. A similar picture is given by the comparison with average German labour productivity growth rates. A faster growth rate in Britain between 1880 and 1900 and a relatively slower one after 1900 combined to give both Britain and Germany almost identical average labour productivity growth rates in the thirty-odd years leading up to the First World War. On the basis of these first findings, British performance, at least with respect to those of its most formidable competitors, looks satisfactory. Given that average French labour productivity growth rates in paper-making over the entire period of this study stood at only 1.9 per cent, Britain's figure of 3.3 per cent looks distinctly respectable.[22]

Table 3.6 *Annual average rates of growth of labour productivity in the UK, US, and Germany, 1860/1–1912*

Year	UK	US	Germany
1860/1–1880/1	1.62	2.26	
1880/1–1900/1	4.46	4.09	3.56
1900/1–1912	1.39	2.20	2.66
1880/1–1912	3.30	3.38	3.31
1860/1–1912	2.65	2.98	

Sources: As for Table 3.5.

[22] This is for the period 1855/64–1905/13. French output is based on data found in T. J. Markovitch, 'L'industrie française de 1789 à 1964 – sources et méthodes', *Cahiers de L'ISEA* AF 4 (1965), Table 1, while labour comes from J. C. Toutain, 'La population de la France de 1700 à 1959', *Cahiers de L'ISEA* AF 3 (1963), Tables 106–110.

Total productivity

As mentioned earlier, labour productivity calculations, however, have their weaknesses; not the least being that they are only partial measures of productivity. An arguably better way of examining how efficiently an industry is using all the resources available to it is total factor productivity (TFP). This technique measures the rate of growth of output not accounted for by the growth of all inputs. The resulting unexplained residual, given certain assumptions, represents productivity growth; or in other words, shifts in the production or cost function. It is thus a catch-all for all factors influencing output other than labour and capital.[23] On the implicit supposition that all opportunities that could be reasonably perceived by local producers were also perceived and used by their foreign competitors, and indeed underlie their faster productivity growth, cross-country comparisons of TFP in the paper industry may go further than labour productivity calculations in suggesting how much potential development British producers were able to realise. It thus allows us to compare the performance of two or more countries whose endowment of capital, labour, and resources may differ greatly. In this chapter, TFP is calculated from the standard Cobb-Douglas generated equation:

$$TFP = \dot{Q} - a\dot{K} - \beta\dot{L} - \psi\dot{R} \qquad (1)$$

where $\dot{Q}$ is the rate of growth of output, $\dot{K}$ the rate of growth of capital, $\dot{L}$ the rate of growth of labour, and $\dot{R}$ the rate of growth of raw materials. α, β, and ψ represent the elasticities of outputs, or assuming competitive conditions, the cost shares of each input. Constant returns to scale is another feature of this particular specification. Assuming that $\dot{R}$ grows as fast as $\dot{Q}$ – which, as we have seen, all the secondary as well as contemporary literature assumes – we can rewrite equation (1) thus:

$$TFP = (1 - \psi)\dot{Q} - a\dot{K} - \beta\dot{L} \qquad (2)$$

Equation (2) is used to estimate the TFP figures in Tables 3.7 and 3.8. Although there may be some doubt as to whether all of the stringent conditions of the Cobb-Douglas function, most notably the assumption of constant returns to scale, actually held in our period, our calculations do at the very least give us some idea of the approximate magnitude and periodicity of productivity change in the British and American paper industries of the late Victorian and Edwardian era. Should increasing

---

[23] M. Brown, *On the Theory and Measurement of Technological Change* (Cambridge, 1966), ch. 7.

rather than constant returns to scale have indeed been the fact in the paper industry, this would have the effect of inflating our measure of TFP by the amount of productivity growth attained through scale economies.

No direct estimates of capital stock in the paper industry for our period are available. Two indirect methods of approximating changes in capital stock are, therefore, employed to calculate $\dot{K}$. $\dot{K}(1)$ uses the stock of paper-machines in the industry, while $\dot{K}$ (2) uses the aggregate width of papermaking machinery. These are reasonable proxies for capital stock in the industry, as the papermaking machine is the key piece of equipment in the mill, and all other capital (e.g. cleaners, boilers, beaters, etc., even buildings) has to expand or contract fairly much in proportion with the paper-machine if the paper-machine is to be fully utilised. Width of machinery is also one the industry's recognised ways of measuring the scale of capital equipment.[24]

Both measures of total productivity given in Table 3.7 trace out a similar productivity history for British papermaking. In all periods significant productivity growth was experienced by the industry, although between 1880 and 1900 extremely high rates were achieved even by earlier standards. These high rates correspond with a technological surge in maximum machine speeds which almost doubled in this period from 300 feet per minute in 1885 to 480 feet per minute in 1900. The American figures from this period exhibit the same trend, but with even higher growth rates. This observation also holds true for the comparison of American and British total productivity after the turn of the century. American productivity growth at all times for which we have

Table 3.7 *Total productivity growth in the UK paper industry, 1861–1912*

Year	$\dot{Q}$	$\dot{L}$	$\dot{K}(1)$	$\dot{K}(2)$	TP(1)	TP(2)
1861/1871	51.04	30.22	20.69	24.64	10.29	9.26
1871/1881	35.86	14.35	26.67	34.92	4.87	2.72
1881/1891	79.19	8.86	−0.56	5.97	28.82	27.12
1891/1901	69.97	16.95	1.51	8.16	22.65	21.16
1901/1907	28.67	18.71	−1.68	−0.22	8.36	8.03
1907/1912	20.60	10.67	−0.38	3.28	6.27	5.45

Notes: Cost shares were for 1861/71–1881/91 $\alpha = 0.26$, $\beta = 0.115$, and $\psi = 0.625$; for 1891–1901 0.224, 0.138, and 0.638, and for 1901/07–1907/12 0.223, 0.160, and 0.617.
Sources: As for Table 3.5, and Weatherill, *One Hundred Years*, appendix C; *Paper Mill Directory*.

[24] A. G. Thompson, *The Scottish Paper Industry till 1860* (Edinburgh, 1981), p. 182.

80   Productivity and performance

Table 3.8 *Comparative US/UK total productivity growth, 1891–1912*

Year	US total productivity	US/UK total productivity
1890/1–1900/1	38.74	171.03
1900/1–1905/7	16.68	199.49
1905/7–1912	8.42	134.36

Notes: K is measured by machine numbers. Cost shares were $\alpha = 0.1961$, $\beta = 0.1021$, and $\psi = 0.7018$.
Sources: As for Table 3.5, and Cohen, 'Economic determination', pp. 5/9, 3/23; *Paper Mill Directory of the World*; Clapperton, *Paper-making Machine*, p. 247.

adequate data clearly exceed British. This finding implies, as did the labour productivity comparison in Table 3.6, that the Americans were outperforming their British counterparts after 1890. Evidence on machine speeds, widths, and capacities presented in chapter 5 also support the existence of this American lead from at least the 1890s. The absence of data prevents a similar comparison with the German industry from being made.

Trade statistics

Leaving direct measures of productivity (and tariffs) aside, one can also get an idea of relative productivity performance from one of its consequences: changing export shares. In Table 3.9 the shares of the total export value of the five biggest producers of paper and Scandinavia (Norway and Sweden) out of the total amount these six exported together are given. Unfortunately, comparable data are not available before 1880. The figures show that after an improvement in the share of exports up to 1888, Britain's share falls from the 1890s, although there is some sign of a slight improvement in 1912. This decline in share in the 1890s coincides with a surge in American and Scandinavian shares. France's share falls quite considerably over the period, while Austria-Hungary's stays fairly constant and Germany's slightly improves in the 1880s and 1890s only to return to its earlier level in the new century. In light of this evidence, despite an annual compound export growth rate of 5.9 per cent, Britain can be seen as losing ground. However, if we consider a larger sample of nations this entirely pessimistic diagnosis may appear premature. Table 3.10 shows from 1888 the shares of total paper exports taken by eleven of the largest producers of the time – together making up at least ninety per cent of world production.[25] With

[25] Mulhall, *Dictionary*, p. 437.

Table 3.9 *Shares of the value of paper exports as a percentage of the total export revenue of six countries, 1880–1912*

YEAR	UK	US	GERMANY	FRANCE	AUSTRIA-HUNGARY	SCANDINAVIA
1880	18.8	4.2	27.3	37.4	11.1	1.1
1882	20.4	5.3	29.9	32.7	10.3	1.3
1884	20.9	2.9	33.5	28.7	12.6	1.3
1886	19.1	3.2	35.7	25.4	15.1	1.5
1888	23.2	2.8	36.2	21.9	13.4	2.5
1890	20.2	3.1	34.6	25.9	12.7	3.5
1892	18.5	3.4	34.1	26.6	12.5	4.9
1894	17.5	4.9	34.4	23.6	12.6	6.8
1896	17.1	5.9	35.3	22.7	11.1	7.9
1898	13.9	12.2	31.5	21.2	11.8	9.4
1900	14.4	11.3	32.4	19.0	12.7	10.2
1902	14.7	13.4	26.2	20.4	12.0	13.4
1904	14.8	12.4	27.4	20.5	11.7	13.2
1906	13.4	12.9	25.6	20.2	12.5	15.4
1908	14.0	10.2	29.7	20.5	10.3	15.3
1910	15.4	9.3	29.9	21.8	8.8	14.7
1912	15.4	9.6	28.2	24.2	7.1	15.6

Source: Statistical Abstract for the Principal and other Foreign Countries.

Table 3.10 *Shares of the value of paper exports as a percentage of the total export revenue of eleven countries, 1888–1912*

YEAR	UK	US	GERMANY	NORWAY	HOLLAND	BELGIUM
1888	17.8	2.2	27.8	0.5	9.9	10.4
1892	14.3	2.6	26.2	1.3	15.7	4.2
1896	12.5	4.3	25.7	2.9	17.7	6.1
1900	10.1	7.9	22.7	2.9	22.2	5.3
1904	10.5	8.8	19.4	3.1	21.9	4.6
1908	9.7	7.1	20.6	3.9	23.0	5.4
1912	11.0	6.8	20.2	4.2	22.3	4.8

YEAR	FRANCE	SWITZER-LAND	SPAIN	SWEDEN	AUSTRIA-HUNGARY
1888	16.8	0.7	2.1	1.4	10.3
1892	20.5	0.5	2.7	2.5	9.6
1896	16.6	0.4	3.0	2.9	8.1
1900	13.3	0.5	2.1	4.2	8.8
1904	14.5	0.6	2.1	6.3	8.3
1908	14.3	0.5	1.7	6.7	7.2
1912	17.3	0.5	0.8	6.9	5.1

Sources: As for Table 3.9.

these figures, Britain's decline as a major paper exporter remains clear, but it is seen to be an experience shared by most of the other old European producers (France, Germany, Belgium, and Austria-Hungary) with the surprising exception of Holland. Although Holland had a significant strawboard industry, its large export shares in this period were due to the fact that a considerable proportion of its recorded exports were in fact the produce of Continental papermakers shipped to their final destinations via Dutch ports.[26] Holland's healthy share of world exports, therefore, undoubtedly greatly overestimates its importance. The clear winners in this period then are Norway, Sweden, and the United States. Seen from this perspective, while Britain's performance in terms of exports was poor with respect to the newer entrants to the industry, who had a remarkably favourable resource endowment, its performance was not all that strikingly dismal when compared with its more traditional competitors whose export markets were likewise falling victim to new blood.[27]

Support for this view is given by information on revealed comparative advantage (RCA) calculated according to equation (3)

$$RCA = (X_{ij}/\sum_j X_{ij})/(\sum_i X_{ij}/\sum_i \sum_j X_{ij}) \qquad (3)$$

where X_{ij} is the volume of exports in industry i in country j. This is a procedure that normalises export shares. By dividing a country's share of world exports in a particular commodity by its share of all world trade, one can identify sectors in which an economy appears to be performing better than average. These sectors have RCA readings greater than one. Although RCA does not actually tell one anything about the underlying sources of such a comparative advantage – whether it be factor endowments, technology, demand, or whatever – it has often been used in the economics literature to support the Heckscher-Ohlin thesis that economies export those commodities that are production-intensive in the factors that are found in relative abundance in the domestic endowment.[28] So, for example, in the twenty years prior to the First World War when wood became the most important source of raw material in the paper industry, one should expect those countries with

---

[26] British Library of Political and Economic Science (hereafter BLPES): Tariff Commission Collection, TC7 28/2 File 167 B(3).
[27] This also appears to be the view of Spicer in Cox (ed.), *British Industries*, pp. 210–11.
[28] B. Balassa, 'Trade liberalisation and 'revealed' comparative advantage', *Manchester School* 33 (1965), 99–123; B. Balassa, ''Revealed' comparative advantage revisited: an analysis of relative export shares of the industrial countries, 1953–71', *Manchester School* 45 (1977), 327–44; N. F. R. Crafts and M. F. Thomas, 'Comparative advantage in UK manufacturing trade, 1910–1935', *Economic Journal* 96 (1986), 629–45; N. F. R. Crafts, 'Revealed comparative advantage in manufacturing, 1899–1950', *Journal of European Economic History* 18 (1989), 127–37.

Table 3.11. *Revealed comparative advantages in paper of eleven producers, 1888–1912*

YEAR	UK	US	NORWAY	GERMANY	HOLLAND	BELGIUM
1888	0.72	0.15	0.79	1.65	1.03	1.99
1890	0.62	0.14	1.26	1.70	1.64	0.99
1892	0.64	0.13	1.99	1.80	1.70	0.77
1894	0.58	0.19	2.77	1.66	1.70	0.99
1896	0.55	0.26	4.04	1.55	1.68	1.09
1898	0.50	0.40	4.78	1.41	2.02	0.71
1900	0.47	0.38	4.39	1.35	2.16	0.94
1902	0.51	0.46	4.15	1.10	2.00	0.82
1904	0.51	0.43	4.77	1.12	1.97	0.78
1906	0.44	0.45	5.77	1.02	2.16	0.86
1908	0.46	0.33	5.93	1.16	2.25	0.96
1910	0.50	0.37	5.60	1.15	2.13	0.71
1912	0.53	0.35	5.43	1.07	2.02	0.71

YEAR	FRANCE	SWITZER-LAND	SPAIN	SWEDEN	AUSTRIA-HUNGARY
1888	1.23	0.26	0.67	0.82	1.60
1890	1.41	0.26	0.77	1.18	1.62
1892	1.50	0.17	0.96	1.38	1.62
1894	1.37	0.17	1.16	1.78	1.36
1896	1.29	0.13	0.81	1.60	1.33
1898	1.27	0.15	0.69	2.02	1.47
1900	1.10	0.20	0.90	2.66	1.50
1902	1.17	0.20	0.83	4.20	1.47
1904	1.21	0.26	0.83	4.04	1.41
1906	1.19	0.21	1.01	4.37	1.58
1908	1.25	0.20	0.82	4.40	1.35
1910	1.23	0.19	0.44	3.75	1.23
1912	1.51	0.20	0.47	3.84	1.05

Sources: As for Table 3.9.

favourable endowments of wood to score an RCA rating of greater than one. The results are given in Table 3.11.

For the entire period for which data are available Britain's RCA readings fall – especially so in the 1890s – showing Britain to be at a growing comparative disadvantage in papermaking. The same experience was shared by Belgium, which moved from a position of strong comparative advantage in 1888 to one of disadvantage in a relatively short space of time. Germany and Austria-Hungary likewise see their strong revealed comparative advantages in papermaking denuded in this period to barely above unity. Indeed, of all the old papermaking nations only the French seemed to have maintained their former advantage, although the scale of that advantage fluctuated wildly. Given France's relatively poor labour productivity performance, this should give grounds for concern that the French, more so than the British, were not utilising their factor endowment to full advantage. The Scandinavian nations' RCA increased markedly over the period. Starting from a position of comparative disadvantage in 1888, both Norway and Sweden acquired a massive revealed comparative advantage of over four by the beginning of the twentieth century. This was the period in which the Scandinavian industry took off. In Sweden, for example, total paper and board output grew in these twenty-odd years from 21,841 tons in 1891 to 215,529 in 1913: nearly a 1000 per cent increase.[29] The United States is the other country whose RCA noticeably strengthens in this period, although it remains at all times in a position of comparative disadvantage. This is a bit surprising given its endowment of raw materials and very strong productivity performance. The explanation lies in the fact that at this time the entire output of the American industry barely catered for its own rapidly growing domestic demand, leaving little production left over for export.[30] In fact, in 1880 only just over 2 per cent of the value of total American production was exported. By 1905, after a period of significant export growth, this figure still amounted to only about 4 per cent.[31]

It must be borne in mind that RCA by itself only assesses where advantages bestowed by resource and skill endowments lie, and *not* performance. The natural temptation to assume that it does gauge performance in some way needs to be resisted. Yet it is worth pointing out that even if RCA is used (erroneously) as a measure of performance, we would find that over the twenty-four years surveyed the RCA of Germany, Belgium, and Austria-Hungary all actually fell more dramatically than that of Britain. Britain's declining RCA in the second half of the century was thus hardly unusual. RCA figures, however, are more interesting if read in conjunction with other factors. Given that British labour productivity figures were comparable or better than the American and German, and that British paper exports grew at an annual compound rate of over 5 per cent in the second half of the century, its low RCA readings start to look less worrying. The argument goes like this: if a nation's productivity performance is good, and it experiences considerable export growth, *even though* its resource endowment puts it

[29] *Sweden as Producer of Wood Goods, Pulp, Paper, Tar and other Forest Products* (Stockholm, 1920), p. 141.
[30] For example, in 1900 American production was equal to only 95 per cent of the domestic consumption of paper in that year. Frickey, *Production*, p. 16.
[31] D. C. Smith, *History of Papermaking in the United States, 1691–1969* (New York, 1970), Table X-3, p. 309.

Table 3.12 *Shares of the value of exports as a percentage of the total paper export revenue of the UK, US, and German Empire, 1872–1879*

Year	UK	US	Germany
1872	46.8	6.2	47.0
1873	47.8	6.0	46.2
1874	42.5	5.9	51.6
1875	43.9	6.3	49.8
1876	41.3	7.0	51.7
1877	39.9	8.2	52.0
1878	38.0	9.3	52.7
1879	32.3	9.6	53.1

Notes: Exchange rates from *Statistical Abstract of the Principal and other Foreign Countries*, except for the $US, which was based on information given in Friedman and Schwartz, *Monetary History*.
Sources: Statistical Abstract of the UK; Smith, *History*, Table X-3, and *Statistisches Jahrbuch für das deutsche Reich* (1880).

at a comparative disadvantage, then it is surely hard to claim that overall performance is bad. Such a combination of occurrences existed in late nineteenth-century Britain. In contrast, a country which has a strong RCA, but still performs badly in terms of productivity and export growth would appear to be one with problems.

The decline in export shares experienced by the British industry presumably began much earlier than the 1880s. Throughout the first half of the nineteenth century, Britain had held the paramount position simply by virtue of the fact that it had been the first nation to mechanise the production of paper successfully. It was only natural then that as continental papermakers began to introduce comparable machinery and bridge the technological gap, Britain's dominance should show signs of weakening. Some idea of Britain's former importance in the international paper markets and the scale of its decline as an exporter of paper is given in Table 3.12. This table gives the share of export revenue of Britain, America, and Germany accruing to each of these nations between 1872 and 1879. It is not possible to go back any further, as before the formation of the German Empire no aggregate figures for output from the various German states are available. Caution, however, should be exercised in taking the German figures for this period as exactly comparable to the British and American figures, for unlike these and the German figures available from 1880, they include *Papier und Pappwaren* (paper and cardboard goods). As a result the German share is larger than it should be probably by as much as 5 per cent. The point to be made from Table 3.12, however, is that as late as 1873, Britain's share of export revenue was still comparable to Germany's. It is probable that if similar data were available for the previous decade, a decade when the British still made more money from exporting paper than it paid out in importing paper, one would undoubtedly find this lead more substantial.

Summary

The purpose of this chapter has been to consider the performance of the British paper industry between 1860 and 1913. In doing this, a wide variety of evidence was consulted, ranging from the observations of contemporaries right through to attempts to estimate by quantitative means productivity growth, export shares, and revealed comparative advantage. Wherever possible British performance has been contrasted with its main competitors in the trade, in this period the United States and Germany. As ever, data availability has been a problem, but as far as it goes the evidence presented indicates that, whilst not brilliant, the British paper industry's performance after 1880 set in its international context was not unusually bad (except perhaps relative to the United States after 1890) and may even have been superior to some of its traditional competitors. This is especially so from the 1880s when Britain increasingly found itself at a comparative disadvantage in the trade; a development attributable to the advent of wood pulp as the industry's chief raw material around this time. This naturally gave great advantage to those producers like Sweden and Norway that were favourably endowed with wood. Earlier losses of export shares were related to the gradual diffusion of the paper-machine to other paper-making countries.

This chapter has raised a number of questions that need to be further investigated. One of these is the industry's choice of raw materials in the second half of the century. It is a fact that until the very end of the century British papermakers used relatively little wood pulp in comparison to their competitors. Was this because of Britain's poor endowment of wood, or was this just another indication of the entrepreneurial failure that is alleged by some to have pervaded late Victorian and Edwardian industry? Another question is that of Germany's export performance. Given the fact that British labour productivity appears at all times to be above German, what explains that country's large share of the world market, and more importantly, Britain's home market? What part did technological and other cost differentials, as well as commercial policies, play in this development? Likewise, the existence of Britain's persistent two-to-one labour productivity gap *vis-à-vis* the United States is a perplexing problem that certainly also warrants further attention. This is particularly so, because all attempts to explain the gap in the literature have been at a highly aggregated level. An industry study of the problem may prove extremely useful. Finally, America's more solid productivity performance and apparent technological surge after 1890 needs to be looked at in more depth to determine if it can tell us what, if anything, was lacking from British papermaking in the years leading up to the First World War. In the following chapters these issues as well as others are considered in more detail.

4 Rags, esparto, and wood: entrepreneurship and the choice of raw materials

One of the most important changes to occur in the paper industry in the latter half of the nineteenth century was its shift to a new source of cellulose. In a trade where over half of all running costs were accounted for by raw materials, it was only natural that considerable attention be directed towards assuring that these raw materials remain cheap and available.[1] This was especially so because, by the middle of the nineteenth century, the supply of the industry's traditional raw material, rag, had shown itself increasing unable to meet the expanding needs of the trade. What followed was an intensive search for a new raw material; a search that spanned the world and which captivated the attention of papermakers for many decades.

At an early stage of this search British papermakers began to focus their interest upon a material that seemed to meet all of their requirements. That material was esparto. Between 1860 and the 1880s British paper increasingly came to be made from this grass that grew in the wild in North Africa and southern Spain. It was a choice of raw material that was peculiarly British, as esparto was not to figure to the same extent, if at all, in the industries of other paper-producing nations of the time. In these nations, whilst esparto was ensconcing itself in Britain, wood was steadily establishing itself as the mainstay of the paper trade.[2] This is an interesting fact. With hindsight we can say with complete confidence that the future of papermaking indubitably lay in wood; an observation confirmed by the fact that from the 1880s British producers did belatedly begin a move to producing paper from wood. As such then, Britain's decision to pursue esparto instead of wood in the 1860s may well have led them down a path that not only undermined the competitiveness of the industry, but afforded its competitors a

[1] Report from the Select Committee on Paper (Export Duty on Rags), *PP* XI (1861), 316. Hereafter *SC Paper*.

[2] 'In Britain, we have taken up esparto to the exclusion of all other new fibres nearly, but on the Continent we find several substitutes have taken a permanent hold of the market ... In France and Germany, they make use of wood fibre, prepared according to M. N. Voelter's patent.' *Paper Trade Review* (hereafter *PTR*), 1 January 1864, p. 1.

valuable technological advantage in the use of a raw material that even British producers eventually realised was the most suitable. Such an erroneous – if that is what it was – decision must surely constitute an example of entrepreneurial failure; yet another disease the British economy is reported by many to have been afflicted with since the Victorian period. This chapter addresses the questions how and why the British paper industry came to adopt esparto as their main alternative to rag in the 1860s with the underlying intention of determining whether this process in any way could provide justification for an entrepreneurial failure thesis in the late nineteenth-century paper industry. Before this can be done, a discussion of the concepts of entrepreneurship and entrepreneurial failure needs to be made. The following two sections of this chapter concern themselves with this.

Entrepreneurial failure

The decline of the British economy since the end of the nineteenth century is a topic that has attracted the concern and consideration of the past few generations of British economic historians. Not without reason, as an understanding of that phenomenon may not only shed much needed light on the current plight of Britain, but also on the dynamics of growth and decline in general. A central element in this debate has revolved around the putative deficiencies of the British entrepreneur. The same entrepreneurship that brought the industrial revolution to Britain at the end of the eighteenth century had been rendered a spent force by the end of the nineteenth; the product of an insipid and vitiating process of gentrification that sapped the very industrial spirit of the nation.[3] The historiography of the entrepreneurial failure thesis should be familiar to any student of the British economy since 1870. As a number of useful surveys are already available in the literature, there is little need here for more than a brief outline of the salient features of that debate.[4]

---

[3] D. S. Landes, *The Unbound Prometheus* (Cambridge, 1969), p. 336. See also, M. J. Wiener, *English Culture and the Decline of the Industrial Spirit, 1850–1980* (London, 1985).

[4] For example, D. N. McCloskey and L. G. Sandberg, 'From damnation to redemption: judgements on the late Victorian entrepreneur', in D. N. McCloskey (ed.), *Enterprise and Trade in Victorian Britain* (London, 1981), pp. 55–72; S. Pollard, *Britain's Prime and Britain's Decline* (London, 1989), ch. 1; F. Crouzet, *The Victorian Economy* (London, 1982), ch. 12; D. C. Coleman and C. MacLeod, 'Attitudes and new techniques: British businessmen, 1800–1950', *Economic History Review*, 2nd series, 39 (1986), 588–611; L. G. Sandberg, 'The entrepreneur and technological change', in R. Floud and D. N. McCloskey (eds.), *The Economic History of Britain since 1700*, 2 vols. (Cambridge, 1981), vol. II, ch. 5.

Early allusions to a belief in a failure in British entrepreneurship are evident in the writings of such luminaries of the late nineteenth- and early twentieth-century world of economics as Marshall, Hobson, Veblen, and Clapham, each of whom seemed to have formed their opinion on this issue on the basis of Britain's poor record in its former staples, coal, steel, and textiles.[5] It was not until the post-war period, however, that questions about British entrepreneurship really came to the fore, most eloquently so in the works of Landes, Aldcroft, and Levine.[6] These works relied almost entirely on qualitative evidence and hoped by collating a great mass of such evidence to establish beyond a reasonable doubt that the failure of the British economy was primarily the result of a human factor – the weakness of British entrepreneurial and technological creativity.

To a new generation of economic historians, trained in economic theory and quantitative methods and eager to use these skills, these arguments by example, which previous economic historians had asked the reader to take on assurance, were unconvincing. In their place they hoped to provide a more rigorous and formal analysis of the hypothesis. For this, it was argued, recourse to neoclassical economics would be helpful in that this would provide a framework with readily testable propositions for the verification or refutation of the entrepreneurial failure thesis.[7] The approach then was essentially one that saw the entrepreneur minimise costs or maximise profits subject to constraints within a comparative static framework. If analysis can show that the decisions of the entrepreneur conform to profit maximisation, then *ex hypothesi*, his or her actions must be considered rational. As the subject matter and evidence of this debate is of an historical nature, the testing of the entrepreneurial failure thesis thus entailed the calculation of the economic consequences of the entrepreneur's decision in terms of the opportunities, or profits, forgone. It is for this reason that this approach has often been called the profitability thesis. In practical terms the test can be carried out in a variety of different ways. It may be expressed equally as the profit forgone, as a productivity differential between nations, firms, or processes, or alternatively, as the distance between production or cost functions. As each method is based on the same

---

[5]   A. Marshall, *Industry and Trade* (London, 1923); J. Hobson, *Incentives in the New Industrial Order* (New York, 1923); T. Veblen, *Imperial Germany and the Industrial Revolution* (New York, 1915); Clapham, *Economic History*.
[6]   Landes, *Prometheus*; D. Aldcroft, 'The entrepreneur and the British economy, 1870–1914', *Economic History Review*, 2nd series, 17 (1964), 113–34; A. Levine, *Industrial Retardation in Britain, 1880–1914* (London, 1967).
[7]   A seminal work in this school was D. N. McCloskey, *Economic Maturity and Entrepreneurial Decline* (Cambridge, MA, 1973).

behavioural postulate, each different test should yield an identical outcome.[8]

Despite the precision and elegance undoubtedly brought to the debate by neoclassical theory, its methods and conclusions have failed to win universal support. One of the most worrying aspects of this approach to many of its critics is that by assuming good information, no uncertainty, and equilibrium as the normal state of affairs, neoclassical economics makes the entrepreneur an everyday manager; passively responding to his environment yet unable to alter it. In this definition of the entrepreneur there is no room for vision, daring, and determination, and the entrepreneurial decision-making process is reduced to nothing more than mathematical calculation. This runs counter to the Schumpeterian idea of the entrepreneur, which distinguishes between the management of the day-to-day running of a firm and entrepreneurship, which is seen as the linchpin of the process of economic development. To those who subscribe to this more dynamic definition of the entrepreneur, the neoclassical challenge to the entrepreneurial failure thesis falls well short of the target.[9] As Lazonick explained, 'the major issue is not, as the new economic historians would have it, whether British management optimised subject to given constraints. Rather, the problem facing British cotton managers was their inability to alter the constraints on investment decisions and profit-making posed by the structure of their industry.'[10] New economic historians were not unaware of this dichotomy of functions, although they obviously ascribed little importance to it in this debate. In an earlier study of the cotton industry Sandberg freely admitted that cotton entrepreneurs devoted 'little or no effort' to developing new techniques and new machinery that fitted their needs, and that this could amount to a 'failure', yet went on to conclude confidently that he had to 'give the entrepreneurs and managers in the industry a relatively good rating'.[11] If the debate has been purely about management, then Sandberg's conclusion is reasonable, and the matter closed. However, if we are discussing the decline of the British economy, or some aspect of it, then there must surely be some consideration of the dynamics of that decline; a consideration which

[8] See P. H. Lindert and K. Trace, 'Yardsticks for Victorian entrepreneurs', in D. N. McCloskey (ed.), *Essays on a Mature Economy: Britain after 1840* (London, 1971), ch. 7.
[9] This, of course, does not mean that management is unimportant or unnecessary, only that it is a separate function from entrepreneurship.
[10] W. Lazonick, 'Competition, specialisation and industrial decline', *Journal of Economic History* 41 (1981), 37. For a similar view, see R. C. Allen, 'Entrepreneurship and technical progress in the Northeast coast pig iron industry: 1850–1913', *Research in Economic History* 6 (1981), 35–71.
[11] L. G. Sandberg, *Lancashire in Decline* (Colombus, 1974), pp. 133, 135.

must take us beyond the myopia of short-run optimisation into a world of uncertainty, chance, and intuition. When we do so we enter the domain of the entrepreneur.

Once this is accepted, the question becomes how can one assess the quality of entrepreneurship, or more precisely within the context of the present investigation, how does one determine whether the adoption of a particular technique or raw material is a sound decision or not? In common with the neoclassical approach, the usual method in the Schumpeterian literature is to make some sort of a comparison. The difference between the approaches, however, is that the Schumpeterian, unlike the neoclassical which compares production functions or engages in a cost-benefit analysis, uses a comparison with a foreign technique not just to test the economic rationality of the actors, but rather to provide an exemplar to which 'correct entrepreneurship' must be heading. For example, in his study of the northeast coast pig iron industry between 1850 and 1913, Robert Allen explicitly used the technological directions being taken in the United States as representing the standards by which the entrepreneurial hypothesis in Britain should be adjudicated. He could thus claim that, 'only the firm that invested in basic steel and American technology was *vigorous*. Those were the firms that were propelling the northeast coast iron and steel industry towards its long-run equilibrium.'[12] Likewise, Lazonick uses American cotton technology and industrial structure as the exemplar to which the British industry should aspire.

Not surprisingly in both instances the industries in question failed the tests. This is only to be expected from an approach that purports to adjudicate entrepreneurship by means of hindsight. To impose a standard that is derived from the historian's own knowledge of the eventual outcome is little more than tautology, for if our decision-rule for the entrepreneurial hypothesis is simply that the successful entrepreneur will adopt the technique that the eventual winner in the contest does, then it is obvious that no one, other than the winner can meet this decision-rule. By definition then, an entrepreneur must always be a winner;[13] and if the winner determines what is good entrepreneurship, by implication this means that there must also only be a single development path for the industry. The course of an industry's progress, however, is unlikely to be so restrictive and deterministic, let alone ethnocentric. Just as with industrialisation, the variants and options open to the entrepreneur and his industry are numerous. Such

---

[12] Allen, 'Entrepreneurship', p. 60.
[13] What is the opposite of entrepreneurial failure? Entrepreneurial success implies infallibility. I prefer the term entrepreneurial competence.

definitions thus impose conditions on entrepreneurship that are every bit as limiting as the neoclassical. Ironically, McCloskey and Sandberg likewise find these ideas implausible:

It is surely driving the theme of the irony of history too far, however, to expect British entrepreneurs to have anticipated the trick history was about to play on them. Indeed, a truly prescient entrepreneur in, say, cotton textiles would have avoided investment in virtually any type of cotton equipment in the years just before 1913, certainly in the very capital intensive automatic looms: if the unforeseeable events of the 1920s and the 1930s are to be made retrospectively foreseeable almost any case of slow adoption of new machinery becomes a rational anticipation of the collapse of Britain's traditional exports.[14]

Their solution was a return to the notion of profit maximisation. But is that the only alternative available? McCloskey and Sandberg come very close to another when they note: 'the issue is what investments in imitation British entrepreneurs could have made that would have been profitable, from their point of view at the time the decisions were made'.[15] If we substitute the word *innovation* for *imitation* in the above quotation, we may just have an appropriate criterion for the entrepreneurial hypothesis. That is: we must consider the entrepreneurial decision-making process *ex ante* rather than *ex post*. Unlike the simple neoclassical world of diminishing returns and profit maximisers where there is only a single solution and the 'best' technology always wins, recent research demonstrates that in the presence of increasing returns and incomplete information there may in fact be multiple equilibria and the possibility of 'inferior' technologies triumphing.[16] Such conditions appear more akin to those confronting the entrepreneur. When the entrepreneur makes decisions he or she stares into the face of uncertainty, but once that decision is made, he/she is committed to it and may not be able to alter it until it is too late. The crucial thing is that no entrepreneur will enter into a venture that he feels will fail. As Kirzner asserts, 'viewed *ex ante*, every entrepreneurial decision taken envisages only profits'.[17] By comprehending the nature of entrepreneurial activity viewed *ex ante*, we are in a better position to judge entrepreneurial competence. The question that must be addressed is not whether in the long run the winning choice was selected, but whether reasonable efforts, or, to use the words that Allen used,[18] sufficient 'vigour' was employed to shape the environment in which the entrepreneur operated. It is indeed grossly unfair to expect the entrepreneur to

[14] McCloskey and Sandberg, 'Damnation', pp. 64–5.   [15] *Ibid.*, p. 65.
[16] Arthur, 'Competing technologies'.
[17] I. M. Kirzner, *Competition and Entrepreneurship* (Chicago, 1973), p. 83.
[18] Allen, 'Entrepreneurship', p. 60.

be always correct; all that can be reasonably expected is that they use their faculties as well as possible. To be sure, this may eventuate in a bet being placed on the wrong horse, but at the time the decision was taken that horse may well have been the favourite, and a bet on the eventual winning horse considered reckless and foolhardy. Moreover, all entrepreneurs must fail at times. After all, not every decision made by a Richard Branson or Rupert Murdoch turns to gold. All that matters for a firm or a nation is that a fair share of successes are achieved.

Entrepreneurial decision-making

If this is to become a workable approach to the problem, the notion of entrepreneurial vigour needs to placed on a firmer basis. Inevitably this requires one to conceptualise the process of entrepreneurial decision-making, as until we have some idea of how an entrepreneur reaches decisions, it is difficult to assess degrees of competence. Perhaps the best place to start is to determine what exactly is the function of the entrepreneur in the workings of the economy. This is a topic that has attracted the comments and opinions of many great minds of the economic world from Cantillon onwards.[19] At a simplistic level the general distinction between the managerial and entrepreneurial function as noted earlier is well accepted in the literature, even if only as a necessary and useful analytical device.[20] Where most dispute has centred has been around whether the entrepreneur's task is to restore or to destroy equilibrium. The latter of the two approaches is the better known and is closely linked to the ideas of Schumpeter, while the former less familiar one is usually tied to Mises. In the Schumpeterian theory of economic development, entrepreneurial activity acts to disrupt the circular flow of an existing state of equilibrium by initiating a process of change. This process, in turn, generates a series of new opportunities that provide the foundation for the establishment of a new equilibrium. The role of the entrepreneur is therefore a disequilibrating one, best encapsulated in the phrase 'creative destruction'.

By contrast, other authors' treatment of the entrepreneur emphasise the equilibrating aspects of his role. The entrepreneur is seen as being crucial in enabling the market process to work itself out in all its contexts. This is achieved by the entrepreneur being alert to all the

---

[19] A good summary of the historiography of this debate is found in R. Hébert and A. Link, *The Entrepreneur* (New York, 1982).

[20] For example, W. J. Baumol, 'Entrepreneurship in economic theory' and H. Leibenstein, 'Entrepreneurship and development', both in *American Economic Review. Papers and Proceedings* 58 (1968), pt. 2, 64, 72–3; A. Cole, *Business Enterprise in its Social Setting* (Cambridge, MA, 1959), p. 12.

unnoticed opportunities and ways of reducing X-inefficiency that render the economy in a virtually perennial state of disequilibrium. With the exploitation of such opportunities the economy is propelled at least temporarily towards a state of equilibrium. In this view entrepreneurship becomes akin to a process of arbitrage.[21] Thus, as a leading proponent of the Misean approach explained, the differences between the Schumpeterian and Misean perception of the entrepreneur is the difference between *causing* cost and revenue curves to shift and being in a position to *notice* that they have in fact shifted.[22] In other words, for the entrepreneur opportunities are noticed rather than made.

Yet, as Hébert and Link have concluded, if we are prepared to accept that there are in fact two faces of entrepreneurship, rather than just the one, then the importance of the debate diminishes considerably. Indeed, from this perspective, rather than representing the antithesis of each other's approach, the Schumpeterian and Misean entrepreneur complement each other. In both views there is agreement that in the wake of each disequilibration, there begins a movement towards equilibrium, although both only offer a partial explanation of this process. Thus, synthesising the Schumpeterian explanation for the cause of the disequilibrium with the Misean explanation for its steady removal, we begin to have a more complete understanding of the process.[23] As a result, the work of the entrepreneur is revealed in all its variety and complexity. An appropriate definition of the entrepreneurial function perforce must capture this diversity. There have been many attempts at definition, although Cole's 'purposeful activity (including an integrated sequence of decisions) of an individual or group of associated individuals, undertaken to initiate, maintain, or aggrandise a profit-orientated business unit for the production or distribution of economic goods and services' seems the best.[24] It is commendable not only for its sufficient generality of function, but also for its realistic allowances for multi-period and multi-party decision-making. But if we are to assess if purposeful activity has taken place we also need to know something about how entrepreneurial decision-making occurs.

If an activity is to be purposeful, then some clear conception of the outcome of that activity and its utility must be in the mind of the entrepreneur when the decision to pursue it is taken. In short, this amounts to saying that an entrepreneur in the performance of his duty

[21] J. A. Schumpeter, *The Theory of Economic Development* (Oxford, 1961), p. 64; L. von Mises, *Profit and Losses* (Sth. Holland, 1951), p. 11. Also see Kirzner, *Competition*, pp. 72–3, and Leibenstein, 'Entrepreneurship'.
[22] Kirzner, *Competition*, p. 81.
[23] Hébert and Link, *Entrepreneur*, pp. 96, 99, 111–14.
[24] Cole, *Business Enterprise*, p. 7.

sets out with a clear intention to break some known constraint or exploit some available opportunity. In other words, the decision-making process begins with an awareness of problems that need to be solved. Kirzner, however, argues that entrepreneurial alertness cannot be deployed to particular problems. Rather, it is a hunch that occurs to the entrepreneur independent of intention or volition: 'entrepreneurship is not an instrument within the decision-maker's grasp, an instrument that he consciously and deliberately deploys in order to achieve an already perceived and desired objective; entrepreneurship is the perception of the worthwhile possibility and the desirability of that objective'.[25] He uses the example of two individuals walking through the suburbs of the same city to illustrate his conception of entrepreneurship. On their journey one notices that there is a difference between the price of apples traded in one part of the city and the price of the same apples traded in another part, while the other is totally unaware of the price differential. To Kirzner the observant one is the entrepreneur, for he has per chance perceived an opportunity to make a gain, and in the process restored equilibrium to the market.[26] Now, while the unpredictable nature of entrepreneurial insight is beyond doubt, and it is plainly fallacious to consider it as just another input in the production function, it is still hard to believe that the extent of that unpredictability is so absolute that there is no scope for the focussing of entrepreneurial attention. There are three weaknesses in Kirzner's example. Firstly, Kirzner's theory has no room for the Schumpeterian entrepreneur, and hence does not consider the removal of constraints intrinsic to the existing equilibrium a legitimate function of the entrepreneur. Once such an entrepreneur is allowed for, the focussing in of entrepreneurial effort onto a particular problem surely becomes more probable. Secondly, as evidenced from the examples he uses, Kirzner's entrepreneur is almost totally devoid of context. He is in a sense an outsider, wandering without purpose or meaning and being randomly bombarded by a multitude of stimuli that may or may not trigger a realisation of a gainful opportunity. While some insights are attained that way, most, we believe, come from those who have some awareness of the aspect of the economy to which the insight pertains. Once again this does not mean that an insider will always be responsible for the introduction of innovations that revolutionise his or her particular industry, or that any insider who puts his mind to it could at any time make the breakthrough, but simply that when the

---

[25] I. M. Kirzner, *Discovery, Capitalism, and Distributive Justice* (Oxford, 1989), pp. 147–8.
[26] I. M. Kirzner, 'The primacy of entrepreneurial discovery', in I. M. Kirzner (ed.), *The Prime Mover of Progress: The Entrepreneur in Capitalism and Socialism* (London, 1980), pp. 10–11, 16–17.

insight is made it is most likely to be made by an insider, because very few people other than an insider would have sufficient opportunity, awareness, and knowledge of the problem to recognise the discovery and the value of its solution. Put in the context of the apple example given by Kirzner, why the two people were walking the streets of that particular city at that particular time, why they took the particular route they did, and what baggage of prior knowledge each was carrying with them, are all surely crucial questions that will influence whether or not an opportunity for profit through arbitrage is perceived. Thirdly, Kirzner's theory and example assumes that a certainty of outcome exists. The observant one in the above example knows for sure that he can make a profit on his apple idea, since he knows for a fact that he can sell his apples at a price greater than he bought them. Entrepreneurial activity, however, is not always characterised by such certainty. Uncertainty of outcome is more likely the norm.[27] When it is introduced an outsider must be less confident of an idea's success and thus less likely to pursue it further. The presence of uncertainty contributes to the greater probability of entrepreneurial insights being made by insiders looking for ways to break down constraints well known to those in the trade.

That entrepreneurial efforts can be generally directed towards certain types of activities by the existence of well-known problems is one thing, but to determine which problem or sets of problems to focus attention on is another matter, about which existing literature has little to say. Most analysis begins with an assumption that the problem to be tackled has been already identified.[28] This is a big assumption to make. Given the complexities of economic life, the frailties of the human mind and body, the influence of institutions and governments, and the laws and vagaries of nature, there exist literally a plethora of constraints for an enterprising spirit to turn itself to. It should be noted that not all of these are necessarily to do with the everyday running of a business. Exerting effort to influence government policy on tariffs, for example, may equally be considered by many an entrepreneur as a legitimate attack on a highly visible constraint. The crucial point is that not all of these constraints can be addressed simultaneously by the entrepreneur. The greater the magnitude of operations which any single individual attempts to direct the less effective in general he or she will be.[29] Implicitly, if not explicitly, this compels the entrepreneur to prioritise the problems that

---

[27] In fact, to some uncertainty is the defining aspect of entrepreneurship. See F. H. Knight, *Risk, Uncertainty and Profit* (Boston, 1921).
[28] For example, M. Casson, *The Entrepreneur: An Economic Theory* (Aldershot, 1991), p. 29.
[29] Knight, *Risk*, p. 282.

he or she feels are soluble: a process shaped as much by cultural as economic factors.[30]

Diminishing returns to entrepreneurial effort, of course, do not only result from spreading attention over too many ends, but may also arise from insufficiently focussing on a specific means of achieving the chosen ends. 'There are many ways to skin a cat' and entrepreneurs just do not have the financial, cognitive, or educational abilities to employ them all. As in Mises' *homo agens*, the entrepreneur is endowed not only with the propensity to pursue goals efficiently once an objective function has been identified, but also the drive and alertness needed to identify which ends to strive for and which means are available.[31] In the end, each entrepreneur must take a judgmental decision, to borrow a phrase from Casson, as to what the targets and decision-rule of his efforts will be.[32] While no hardfast rule can be set down for such a subjective process, one could reasonably expect past experience, expectations, the amount of information, and cultural prejudices to play important roles.[33] One should also expect that these judgmental decisions are not immutable. In entrepreneurial decision-making that is multi-period, information and expectations are in a constant state of flux and may bring changes that, if compelling enough, lead to minor and sometimes possibly even radical alterations in the original definition and priorities of ends and means. The incorporation of new information and re-defining of objectives is an absolutely essential aspect of the entrepreneurial decision-making process. Without it entrepreneurship would lose much of its cutting edge.

This alteration of ends and means also highlights another aspect of entrepreneurial decision-making: namely, that a good deal of it is of an on-going nature. This is in contrast to the frequently used assumption that the process of entrepreneurial decision-making can be adequately understood through single-period analysis. Rarely would the entrepreneur's work consist solely of a 'once-and-for-all' decision. Rather, most projects are more likely to involve a continuing participation by the entrepreneur in which the advent of new information and circumstances generates a sequence of continuously changing decisions. In this view, information, especially its availability and its accessibility, becomes of vital importance. Moreover, the completeness and reliability of available information plays a part. At the outset of an entrepreneurial endeavour information may be patchy, of poor quality, misleading, or non-existent,

[30] M. Casson, 'Entrepreneurship and business culture', Discussion Paper in Economics, University of Reading, no. 239 (1991), p. 29.
[31] L. von Mises, *Human Action: A Treatise on Economics* (New Haven, 1949).
[32] Casson, *Entrepreneur*, p. 29.   [33] Cole, *Business Enterprise*, pp. 26–7.

and the scope for uncertainty great. In such circumstances, while most of us would probably settle for the status quo, action is taken and decisions made by entrepreneurs. Reasons for this go to the very heart of human creativity and rarely can be expressed in anything other than vague terms such as animal spirit, instinct, drive, purpose, or self-righteousness. As Casson put it, 'the entrepreneur believes he is right, while everyone else is wrong'.[34] The apparent intractability of such attributes to rigorous analysis may cause some theorists to question the value of these qualities, but it does not make them any less real. Their importance, however, would seem to diminish over time as more information on the problem facing the entrepreneur comes to light. This information comes from a variety of sources: prices, technological and scientific developments, changes in the political and legal environment, industrial espionage to name but a few.

A very important source, particularly when the decision has involved a total step into the dark, is the firm's own decisions made at prior stages of the project. Starting from a state of complete ignorance about likely outcomes, the implementation of the initial decision to pursue that particular course necessarily will engender the creation of germane information, so that when the decision is reviewed again at a later date reconsideration can take place in light of this new information. In this manner a continuous stream of data, on the basis of which future decisions are formulated, is monitored by the entrepreneur. But these data are often of a very specific nature in that they can be heavily weighted towards the original option pursued by the firm. This is perfectly understandable, because if the firm experiments with method X, then unless there is an alternative method Y very similar in character to X, the information generated by the experiment is highly unlikely to tell us anything about methods other than X. Except when the information attained suggests that method X is not feasible, this should create a bias in future considerations towards a continued perseverance, albeit with possible modifications, of method X. This bias is also augmented by the creation of physical network externalities of method X. Investment in plant, machinery, training, as well as commercial and marketing relations for method X, can weigh heavily in favour of X, irrespective of the superiority, or otherwise, of the method.[35]

All of this is relevant information to the entrepreneur. But entrepreneurs need not be inevitably locked-in to a particular path by their early decisions, for rarely if ever is an entrepreneur or firm alone in exploring a potentially fruitful innovation. The problems and constraints that he

[34] Casson, *Entrepreneur*, p. 14.
[35] David, '*Clio*'; Arthur, 'Competing technologies'.

must confront daily, and possibly even the vision of the future of his industry that he holds, must be shared to some extent by others in his field. In as far as they also look for directions and solutions, their findings, as do impersonal changing market signals, also impinge on the entrepreneur's future decisions in these areas either reassuring the entrepreneur of the rectitude of his hunch or convincing him or her that it needs to be modified or even abandoned. The demonstration effect can thus be a very potent tonic to the tendency of path-dependency.

Entrepreneurs acquire much of their knowledge and information from their day-to-day experiences not only in the pursuit of their own businesses, but in living in a vibrant and transforming environment. These experiences provide inducement to reflection, reinterpretations, discoveries, and generalisations.[36] This steady accretion of pertinent information gives rise to the ever-evolving nature of decision-making. If decision-making is about information and its processing, then it is surely reasonable to assume that the quantity, quality, and nature of that information, as well as how it comes to the entrepreneur, should directly influence the shape and method of that decision-making. As time and experimentation progresses, it is to be expected that uncertainty, and with it the role of intuition in the decision-making process should diminish. In its place more rationally based modes of calculation, ever more closely approximating firstly the model of decision-making posited by bounded rationality,[37] and later neoclassical economics, come to the fore. In time then, the entrepreneurial decision characterised by uncertainty, gut feeling, and caprice eventually collapses down to a problem of efficient management. To be sure, the time that it takes to become purely a problem of sound management varies considerably from case to case, and in some instances, where the flow of vital information is negligible, it may be hard to conceive of it ever reaching such an outcome. However, the main point is that entrepreneurial decision-making is an evolutionary process, initiated by instinct but increasingly influenced and transmogrified by information. In other words, it may be possible to explain the pattern of change in the entrepreneur's decisions as the outcome of a learning process generated by the unfolding experience of the decisions themselves.

It might be argued that this model of entrepreneurial decision-making with its emphasis on the gradual revelation of information over time provides strong justification for the 'fast second' approach to entrepre-

---

[36] F. Machlup, *Knowledge and Knowledge Production* (Princeton, 1980), p. 179; Kirzner, *Discovery, passim.*
[37] For more details on bounded rationality, consult H. Simon, *Models of Bounded Rationality* (Cambridge, MA, 1982).

neurship. This is the approach that states that it is rational for entrepreneurs to hold back on making crucial decisions, at least until others have worked out better the ramifications of all of the alternatives available. Interestingly, this is the direction recent research in investment theory has gone.[38] Given the impossibility of entrepreneurs devoting equal attention to all the potential solutions to a problem, it does make a lot of sense for the entrepreneur to adopt a wait-and-see approach for those solutions that he chooses not to pursue actively. But this is not what is exactly meant by the 'fast second' approach. Moreover, the type of decision we are concerned with here differs from that to do with the everyday investment in capital stock that is usually considered in the literature. In more dynamic surroundings waiting may prove very costly indeed. By sitting back and allowing others to pioneer new fields and markets, a firm not only forgoes current profits, but more importantly, may actually place itself in an untenable position in the struggle for survival in a changing and competitive environment. Although the waiting firm can expect the discovery of relevant information to be a function of time, it has no guarantee that this information will be promptly, if at all, received; or that if it is, it will actually be able to implement or even understand it without first familiarising itself to the same extent as the discovering firm with the intricacies of the solution. Taking these factors into consideration, failure to be at, or near, the cutting edge, rather than being a clever strategy, places the firm in a situation where it always runs the risk of lagging behind its competitors or even of being eradicated long before it ever gets the chance to catch up. 'Theories of optimal inertia' are thus no substitute for entrepreneurship.

In its emphasis on information this approach is similar to neoclassical decision-making. The difference, however, comes from its consideration of the creation and impact of new information – much of it itself a product of prior decisions – on the decision-making process. Neoclassical analysis when it considers a long-term shift to new methods does so within a framework of a long-run equilibrium disturbed by factors exogenous to the system.[39] These factors, usually technological, manifest themselves in relative prices and bring about a substitution towards the now relatively more efficient method; a process which restores equilibrium. The diffusion of a new method thus is mapped as a series of

---

[38] A. Dixit, 'Investment and hysteresis', *Journal of Economic Perspectives* 6 (1992), 107–32; and R. S. Pindyck, 'Irreversibility, uncertainty, and investment', *Journal of Economic Literature* 29 (1991), 1110–48.

[39] For example, see C. K. Harley, 'The shift from sailing ships to steamships, 1850–1890: a study in technological change and its diffusion', in D. N. McCloskey (ed.), *Essays on a Mature Economy* (London, 1971), pp. 215–34.

shifting rival production functions with the balance determined at any point by the reigning market conditions of the time. Crucial to the process is the assumption of the availability at all times of adequate information about the alternatives laid before the manager. In contrast, with the decision-making process as described above, this assumption is only likely to hold towards the end of the project when sufficient technological and market evidence has been collated. Before we can get to that stage of comparing two or more fairly well-established methods, we should surely know something about how the competing methods arose. Far from being manna from heaven, each in fact represents the culmination of the efforts of numerous inventors and entrepreneurs who were prepared to take a chance on an unknown. To exclude this from a consideration of the entrepreneurial merits of a decision is to ignore significant aspects of the history of the choice and effectively to bypass the entrepreneur himself. At earlier stages modelling decision-making behaviour along neoclassical lines is inappropriate, failing as it does to capture the uncertainty and serendipity of the process as well as the gamble that is the hallmark of entrepreneurship.

There are also many similarities with Nelson and Winter's evolutionary approach to the firm.[40] It differs, however, in its consideration of the decision-making process. In Nelson and Winter's model, the firm, unable to profit-maximise because of bounded rationality, is compelled to satisfice: that is, to continue in its routine behaviour as long as a specified rate of return is achieved. This decision-rule in their model is taken as a constant. By contrast, in the approach outlined above, the implications of bounded rationality as well as path-dependency are applied to the determination of the decision-rule itself. These are portrayed as being shaped by the flow of new information. Nelson and Winter's satisficing behaviour, moreover, is applied equally to entrepreneurial as well as managerial decisions, even though it seems to fit more comfortably with the managerial role than with the entrepreneurial. Even the choice of the word for what the firm does, 'routine', belies any entrepreneurial function it might contain. This is also seen in the fact that in their theory failure to achieve the satisficing level is said to induce a process of search for new routines or to imitate the successful routines of its competitors. This process of search approximates more closely to the activity of entrepreneurs. However, we are told little about this process, except that it is local and stochastic. This is an unsatisfactory account of entrepreneurship. The aim of entrepreneurship is not to continue in one's comfortable ways, but to challenge and improve upon

---

[40] Nelson and Winter, *Evolutionary Theory*.

them. To argue then that entrepreneurship is essentially a reactive force understates the importance of this factor. A fuller investigation of the process of search would show its importance for decision-rules, as the degree of boundedness the firm faces must surely be dependent on the success (measured in terms of information acquired) of such searches. It would also suggest for similar reasons that each aspect of the firm's activity that requires decisions to be made may not be guided by the same decision-rule, and if they are, this rule may not be rationally based. In terms of our framework, Nelson and Winter's model, like the neoclassical, represents one historical stage in the decision-making process: a middle stage where there is adequate information for bounded rationality, but insufficient information for perfect rationality. It is too inflexible to tell us all we need to know about entrepreneurial decision-making.

What are the implications of this section for the study of the entrepreneurial failure hypothesis? For one thing it indicates that what we really should be expecting from British entrepreneurs is competence, not omniscience. Competent entrepreneurship is about the exertion of sufficient vigour in an assault on a constraint, not the infallibility of these efforts. We must be more specific. The level of vigour alone is in itself insufficient, as the vigour must also show purpose. To do this the efforts of the entrepreneur must reflect the circumstances in which the varying phases of a decision take place. This is something for which it is difficult to get a convenient measure. R&D expenditure and patents may provide possibilities, but these must be used with the greatest of care, since they tell us little about the context and content of such exertions. Moreover, their use is compounded by the restrictive paucity of such data available for historical studies. Comparative productivity performance across countries, such as those analysed in chapter 3, may also be of use. This is not because any other country's performance necessarily represents the 'ideal', but because one could expect a fairly equal distribution of successes across countries. A significantly differing rate of productivity growth would in this case suggest a failure of some sort. Unfortunately, productivity figures alone do not help us locate the source of the relatively poor performance. Under such conditions it must be left up to the historian in each particular case to scour the primary sources to find appropriate means for ascertaining the presence or absence of sufficient vigour. For the historian this would involve an analysis from the perspective of the entrepreneur of the entire history of the decision as it unfolds from the formulation of the problem, through its initial implementation and subsequent development, until its eventual denouement. This must be done with knowledge of how prior decisions and the inflow of information bear on this process. It is only with the appropriate consideration of the evolution of the problem and the choices presented to the entrepreneur that the vigour with which these options are pursued – and hence the soundness of entrepreneurship – can be adjudicated. In the following sections this approach is applied to the British paper-maker's choice of raw materials in the second half of the nineteenth century.

The supply of rag

Ever since rag replaced parchment as the chief source of cellulose for the British industry sometime in the late Middle Ages,[41] its availability and supply price have been matters of great concern to the British paper-maker. The perennial fear was that the supply of rag would be unable to grow apace with the ever increasing demand for paper. The first recorded suggestion in the West that an alternative material to rag could be used for papermaking was made by Edward Lloyd, a don at Jesus College, Oxford in June 1684. His chosen substitute was asbestos, and although Lloyd's proposal was never taken seriously by papermakers, it did spark an interest in others.[42] Not too long after that the celebrated French naturalist and physicist, René Antoine Ferchault de Réaumur, delivered a treatise on the subject to the French Royal Academy. In this treatise, Réaumur proclaimed that paper could be made from wood; a conclusion he had reached after observing the ability of wasps to convert wood filaments into a paper-like substance used in the construction of their nests. The motivation for Réaumur's discovery was apparent:

This study should not be neglected, for it is, I dare say, important. The rags from which we make our paper are not an economical material and every paper-maker knows that this substance is becoming rare. While the consumption of paper increases every day, the production of linen remains about the same. In addition to this the foreign mills draw upon us for material. The wasp seems to teach us a means of overcoming these difficulties.[43]

Many others followed in the footsteps of Réaumur, expanding the number of known substances which could be used by papermakers. Between 1765 and 1771, Dr Christian Schäffer, a clergyman in Regensberg, Bavaria, with a passion for botany and natural history, compiled a six-volume work dealing with the use of vegetable matters in papermaking. This monumental work, whose influence in the scientific

[41] Hills, *Papermaking*, p. 2.
[42] D. Hunter, *Papermaking: The History and Technique of an Ancient Craft* (New York, 1974), pp. 311–12.
[43] *Ibid.*, p. 314. The treatise appeared in 1719.

community spread even to Britain, identified hundreds of species from which a reasonable paper could be fabricated. At the beginning of the nineteenth century Matthias Koops began experiments in London which resulted in him taking out patents for processes that used wood and straw, as well as one that de-inked paper. He wrote three books on these matters, which were printed upon paper made with his processes. To many paper historians Koops' work represents the beginning of the modern paper industry, as his contributions laid the groundwork for, and greatly influenced, Friedrich Gottlob Keller's invention of the first working wood-grinding machine around 1840.[44]

Since the beginning of the eighteenth century scientists and observers have thus been turning their minds to discovering a substitute for rag. When this became the burning issue in the 1850s, what was new was not the fear of insufficient rag supplies, but a belief that its realisation was rapidly approaching. This belief, in turn, emanated from misgivings about the upward trend in rag prices. Although rag prices had on a number of occasions soared to great heights, such as during the Napoleonic Wars, these had been viewed as temporary fluctuations that, once normalcy had been restored, would pose no threat to the long-term viability of the industry. With the 1850s, however, began an acceleration in this upward trend of rag prices, which unlike previous upswings showed no apparent sign of abating and which outstripped the general rise in the price level of the time. Although the inadequacies of the data prevent the construction of a price series, the fragmentary evidence available does provide us with some illustration of the extent of the rise in price. The price of rags used in making ordinary printing-paper, for example, rose by 28 per cent from an average of 12s 6d per cwt in 1848–52, to 16s per cwt in 1853–6. Most other grades showed similar increases of 20 to 30 per cent between 1852 and 1860.[45] The average price of imported rag also increased from about £20 per ton in 1853 to around £23 in 1858, and by 1861 a typical German rag cost £30 in Britain.[46]

The sources of these price rises were manifold. The demand for paper over the century, and especially from the mid-century, grew prodigiously with *inter alia* the spread of the penny press, packaging, and better postal services. As a result the demand for rag grew faster than its supply. Factors independent of the paper industry contributed to this failure of the supply of rag to keep up with demand. One key source of rag was the waste product from Britain's cotton and linen factories.[47] Despite the

[44] For a brief history of these early experiments, see, *ibid.*, ch. 11.
[45] Coleman, *British Paper Industry*, pp. 214, 338.
[46] Spicer, *Paper Trade*, diagram III, p. 32; *SC Paper*, p. 284.   [47] *SC Paper*, p. 316.

fact that these industries grew at a fast rate in this period, their growth did not result in greater quantities of waste for the papermaker, as improvements in weaving technology, as well as the advent of alternative uses for this waste, greatly reduced the amount of it left over for papermaking. According to Frederick Magnay in 1861, 'cotton waste was first of all given away; it was then £2 or £3 a ton, and now something like £8 a ton'.[48] Chief among its alternative uses was as a cloth for cleaning equipment and machinery particularly in the railway and shipping industry. Of course, once the material had been used for this purpose and was saturated with grease and oil, its value to the papermaker was gone.[49] Changes in the taste of consumers from cheap cotton and mixed cotton and linen fabrics to cheaper woollen and mixed woollen garments less suitable for papermaking were also blamed for the dwindling supplies of rag.[50] Moreover, another destination for used rags, the Old Clothes Exchanges which bought old clothes, cleaned, repaired, and then sold them to clothing merchants in Ireland, Belgium, and Holland, competed strongly with paper-mills for used rags. There were some two to three miles of these old clothes shops and stalls in and around Petticoat Lane in London alone.[51]

International factors also played a part, especially after the repeal of custom duties on the export of rags. This enabled foreign producers, particularly the Americans, to get free access to Britain's rag supplies, and as the exports of these materials picked up over the 1860s from around 1,000 tons in 1861 to 24,000 tons in 1870, this in turn bid up the market price of rag in Britain.[52] The other papermaking nations failed to reply in kind to Britain's liberalisation of her rag market and maintained, indeed in some cases increased, their existing export duties. In France the export duty in 1861 stood at £4 17s 2d per ton; in Germany £9 3s; while in Belgium the export of rag was totally prohibited.[53] As Britain continued to need to import rags from the continent to meet its requirements – somewhere between 9 and 20 per cent of its total rag needs – these foreign export duties forced the price at which rag was available to British producers up by, according to John

[48] *Ibid.*, p. 294.
[49] Copy of a correspondence between the Departments of the Treasury and Board of Trade, in regard to the increasing scarcity of the materials for the fabrication of paper, *PP* LXV (1854) (hereafter *Correspondence*), 495.
[50] *SC Paper*, p. 316.
[51] R. Turvey, 'Economic growth and domestic rubbish, London 1855–1926', paper presented at the ESRC Conference on Quantitative Economic History at St Antony's College, Oxford, September 1992, p. 9.
[52] Coleman, *British Paper Industry*, appendix IV.   [53] *SC Paper*, p. 269.

Evans' estimate, as much as a third.[54] Producers understandably viewed their future with pessimism.

One course of action open to them was to try to enlarge the home supply of rag. Rag collection, however, was a trade totally independent of papermaking, so that the papermaker had little scope to influence the supply of the raw material. The rag trade was usually organised by wealthy rag merchants. In late Victorian Britain the market for recycled waste products had already developed quite a deal of specialisation. Rag-and-bottle stores and Marine stores were exceedingly numerous in the poorer sections of London, and each type of store was careful not to deal in the merchandise of the other. The chief distinction between the two types of stores appears to have been that the rag merchant concentrated on materials that could be sold as inputs in the manufacture of some new product, whereas the marine-store man usually only handled rubbish that could be resold directly to the consumer for the purposes for which they were originally intended.[55] The rag merchant relied upon a legion of rag-and-bone men to go around the towns and cities of the region collecting any rags they could lay their hands on. This they did, often to the chagrin of the passerby. In 1872, for example, a complaint was recorded in Westminster about several rag-and-bone men who had disrupted the free movement of people and goods in Great George Street by offloading the contents of two dust trucks onto the pavement to sort through for rags, bones, and paper. More often than not these rag-and-bone men were not in the direct employ of the rag merchant, but were paid, often in kind, by the weight and quality of the rag they brought to him. The only persons the merchant actually employed were those who helped him co-ordinate the business and negotiate with paper-mills and regular suppliers such as cotton manufacturers. If the firm was large enough, women might also be employed to sort the rags according to recognised grades and prepare them for sale.[56] An example of such a business was that run by William Petty in Edinburgh in the second half the nineteenth century. On his premises in Holyrood Square, Petty ran a warehouse that usually held around 70 tons of rag at any one time and where he employed a 100 workmen.[57] This method of organisation made sense, since the collection costs of rags in small quantities from many sources made the direct involvement of the rag merchant in the collection of the rags unprofitable. Occasionally the rag-and-bone man might deal directly with the papermaker, although this

[54] Coleman, *British Paper Industry*, p. 214; *SC Paper*, p. 283.
[55] Turvey, 'Economic growth', pp. 8–9.
[56] Spicer, *Paper Trade*, pp. 9–10; Coleman, *British Paper Industry*, pp. 37–8, 166–7, 338.
[57] *PTR*, 26 September 1884, p. 196.

only occurred in rural areas where no rag merchant had established himself. Another source of rags for the rag merchant was the dustman of late Victorian British cities. Dust collection in these cities was predominantly organised by local vestries and district boards which in the vast majority of cases farmed the task out to private contractors. Amongst the rubbish amassed every day, a collector could always count on finding a certain amount of recyclable material that was worth money to him. Upon arrival at the dustyards then, all rubbish was gone through, and valuable material such as rag set aside for sale to the rag merchant.[58]

A number of people at the time, usually not directly involved in the trade, felt that with a degree of better organisation the supply of rag could be sufficiently augmented to allay the papermaker's fears. Despite the protestations of the papermakers present, Dobson Collet, secretary of the Association for Promoting the Repeal of the Taxes on Knowledge, could confidently announce to the Select Committee on Paper in 1861 that, far from there being a shortage, there was in fact an ample supply of rag in Britain to meet the growing needs of the paper industry. This sanguinity was based on an inquiry allegedly made into the matter by several of his colleagues in Birmingham which suggested that as little as 40 per cent of that city's supply of rag had been tapped by existing collection mechanisms. The attainment of a sufficient supply of rags was thus merely a matter of better planning; planning which he thought would be induced in a short space of time by the prevailing high prices in the trade.[59] Richard Herring, a man who, unlike Dobson Collet was not unfamiliar with the trade, argued in 1860 more specifically for the need for better organisation:

Collection of rags has hitherto been by a small trap, and in the hands of petty dealers; the general carelessness of collection and lowness of price have equally diminished the quantity. It has been ascertained that in scarcely fifty houses out of every hundred is a collection made, and the negligence arises partly from mistakes as to the nature, value, and manner of the due collection.
... Every housewife ought to have three bags; a white one for the white rags, a green one for the coloured, and a black one for the waste paper (the three might be furnished for a shilling), and these would prevent litter, waste, and the trouble of collecting when the demand came. A suitable agency formed in the towns and villages would settle all demands, arrange the contributions, and reduce the whole into a regular trade. The general apprehension that we require French or foreign rags for our manufacture is a mistake; we have a sufficient supply at home if we will but make use of it. There are more rags wasted, burnt, or left to rot than would make our papermakers independent of all assistance from abroad. A regular communication ought to be formed by country carriage and by railroads for the conveyance of the bags to London, or to those mills in the

[58]   Turvey, 'Economic growth', p. 13.   [59] *SC Paper*, p. 376.

country which enter largely into the trade. A plan is proposed which will place the whole subject plainly before the public, offer proper pledges, establish proper means, and give the whole movement the degree of activity and regularity which may render it profitable to individuals and the country. A little industry, a little intelligence, and an established system would perfectly secure us from failure in an important branch of art and trade, already worth six millions sterling, employing a large number of skilled workmen, and conducting most effectually to the industry and comfort of the peasantry and to the trade of the Empire.[60]

Eyes were also cast across the Channel to the efficacy of continental means of acquiring rags. The *Paper Trade Review* in 1864 spoke in glowing terms of the 'Rag Pickers of Paris', the army of 25,000 youngsters which scoured the streets of the city for suitable rags. Of this number, a good 15,000 of them were juvenile offenders, so that the operation was run with almost military precision by the prefect of police. The pickers were paid by the weight they collected, and the going rate in the early 1860s was 2.5 francs per basket. According to the report, 62,500 francs worth of rags were collected in the Parisian area daily. Similar schemes were under way all over France, giving employment to over 100,000 people.[61] Similar efforts were attempted in the London area, though never on anything like the scale of the Parisian project. The first report of a rag-collecting brigade in London appears in 1863.[62] Established in the summer of 1862 under the patronage of the Earl of Shaftesbury and managed with business-like efficiency by J. H. Lloyd, the 'Ragged School Union' as it came to be known, operated by assigning three boys to each of the organisation's seven trucks, instructing them to perambulate through assigned districts and collect paper, rags, bones, old metal, and any other potentially valuable rubbish. In its first year of operation it made a modest profit of £240, and on the basis of that success a call went out to extend the organisation to other parts of London as well as to other cities.[63] These plans seemingly never came to fruition, and the collection of rags in Britain remained predominantly in the hands of the rag merchant. In these hands many contemporaries were openly sceptical about the chances of ever seeing a greater supply of rag. As Dr Forbes of East India House wrote in a report to the Board of Trade in May 1854, rag merchants were highly likely for a time 'to influence the market, both as to supply and price, by withholding their stocks'.[64] The willingness of the rag merchant to assert what market power he had was obviously also well known to the papermaker, who told a governmental committee in 1804:

[60] R. Herring, *Paper and Paper Making, Ancient and Modern* (London, 1863), p. 71.
[61] *PTR*, 1 April, 1864, p. 100.   [62] *PTR*, 1 August, 1863, p. 160.
[63] *PTR*, 1 December,1864, p. 283.   [64] *Correspondence*, p. 495.

'the Rag Merchants, I consider (between ourselves) d—d deep, low, cunning, and cheating Chaps, who don't care who is ruin'd, if they get £5 by an advance'.[65] Moreover, the fact that rag was only one of the many items that the rag merchant actually collected meant that the dampening of prices of these other goods could have offset any incentive given by higher rag prices. To some extent this was also reflected in the fact that master dustmen, who earlier had submitted tenders to vestries to have the liberty of collecting the rubbish, in the second half of the nineteenth century began to demand remuneration for the same efforts.[66] Still, the need for some dramatic response to the scarcity of rags was to prove academic. With the appearance of new raw materials on the scene in the coming decades interest in and a perceived need for rag-collecting brigades and other methods of enhancing the supply of rag disappeared.

Even so, all the efforts that had already been made, as well as those that could have been made, had they been necessary, at best would only have had an exiguous effect on the ever-widening gap between the demand and supply of rag that had and would have continued to have opened up. The exhaustion of the supply of rag is suggested by the fact that in the vestry of St Luke in London in 1882/3, for each 1000 cart loads of rubbish, weighing some 1,250 tons, collected in that parish in that year, a mere 0.84 tons of rags were recovered. As a source of recycled rubbish, this figure put it well behind glass, iron, bones, paper, and bottles in importance.[67] On this matter, therefore, it seems safe to concur with Spicer's finding that as long as rag remained the primary source of cellulose for papermaking 'no amount of economy and care in rags would render the total supply in any way adequate to the modern need for paper'.[68] If the supply of rag could not be adequately enhanced, due to its inherent inelasticity, then alternative solutions had to be found. One possibility would have been to improve the chemical processing of rags, so that rags that had previously been unsuitable for papermaking would now become available. This avenue was certainly pursued, but was never likely to produce an increase in supply that would have satisfied the voracity of the papermaker. In any case, the introduction of chlorine bleaching in the last decade of the eighteenth century had already opened up the world of the non-white, soiled rag to the eager papermaker, leaving relatively few untapped sources of rag available.[69]

---

[65] Coleman, *British Paper Industry*, p. 276.
[66] Turvey, 'Economic growth', pp. 14–15.   [67] *Ibid.*, p. 14.
[68] Spicer, *Paper Trade*, p. 29.
[69] Coleman, *British Paper Industry*, pp. 113–16. Slight improvements, nonetheless, were made on this front. See, e.g., *PTR*, 1 September 1863, p. 177.

Efforts therefore largely turned to finding a substitute for rag. As Schäffer had ably demonstrated almost a century earlier, there was no shortage of candidates. The challenge was to find one (or more) that could meet the papermaker's notions of commercial viability. 'I am quite sure', the Scottish papermaker Henry Bruce told the Select Committee on Paper in 1861, 'there are very few fibrous plants between Land's End and John-O'-Groat's house that might not be made into paper; but I am also quite sure that there is not one as cheap, but every one a great deal dearer than ever dear rags.'[70] The search continued, nevertheless, and hundreds of ingenious, if sometimes bizarre, attempts were made to find a solution to the problem. Perhaps the most amazing solution was that of Augustus Stanwood, a New England papermaker, who imported mummies from Egypt for the sole purpose of stripping the corpses of their cloth wrappings and using the material for making coarse brown wrapping paper.[71] In the following sections I will look at the search for new papermaking materials in Britain; the costs of the different raw materials in the early 1860s; a case study of the transition from rag to esparto to wood in one firm; and finally the same transition in the industry as a whole in the latter half of the century.

The search for new raw materials

The threat of dwindling supplies of raw materials, that, especially from the mid-century, hung like a sword of Damocles over the heads of British papermakers, acted as a powerful stimulant to the search for a new source of cellulose. It soon began to have its effect. In a particularly optimistic tone the *Paper Trade Review* in 1863 told its readers:

We have experienced at the early part of the year a great demand for materials in almost every quarter of the world, and in many places the material was completely exhausted. This at once gave a power to the other fibres in the market, and at no former time, in the state of the trade was the papermaker so willing to make use of other materials in lieu of rags ... The dearness and scarcity of rags has, in some cases, compelled the makers to experiment and make use of other substitutes in the room of rags.[72]

Many papermakers exerted not inconsiderable effort in the search. Frederick Magney of Delare and Magney and Co. assured the Select Committee on Paper that they were 'constantly making experiments, either as to raw material or as to the colour of the paper, to improve it'.[73] Moreover, no pains or costs were spared in the process. This was not unusual behaviour; most papermakers before the same Select

[70] *SC Paper*, p. 304.   [71] Hunter, *Papermaking*, pp. 382–5.
[72] *PTR*, 1 January 1864, p. 1; and 1 August 1863, p. 154.   [73] *SC Paper*, p. 296.

Committee also testified if not to having actually invested similar amounts of time and money in experiments, then at least to having been *au fait* with the current developments in the area. John Carlise, managing partner of C. E. and J. G. Potter of Darwen, Lancashire claimed to have already spent £3,000 to £4,000 on testing out new fibres; the owner of Roughway Mill in Kent had made extensive practical trials with beetroot, common matting, esparto, wood, and straw; Henry Bruce, the Scottish papermaker, had sought and received specimens of fibrous plants from the West Indies, the Subcontinent and America; while William Tullis of Tullis and Co. had travelled to see the new materials himself, particularly spending a good deal of time in Spain enquiring about esparto.[74] These examples could be replicated a number of times over. The point being made here is that in the mid-nineteenth-century paper industry it was exceedingly common for papermakers to try out new materials as they became available. There was certainly no lack of vigour on this score.

The lone dissenting voice at the hearings was Collet Dobson, the representative of the Association for Promoting the Repeal of the Taxes on Knowledge. He accused the industry of conservatism and complacency in the adoption of new materials: 'if a man is doing very well and is on the wrong side of fifty or sixty, he is not always disposed to make a change'.[75] This was part of his attempt to undermine the papermaker's appeal for the restoration of export duties on rags. It will be recalled from earlier in the chapter that he also believed there were sufficient supplies of rag in Britain as well. Dobson's charge, however, was not only tendentious, but erroneous, based as it was on a belief that all that was needed was a vegetable fibre that could be transformed into paper, when in fact it had also to meet stringent cost and quality requirements. Wrigley, back before the Select Committee to give further testimony, took umbrage at this suggestion that papermakers would shy away from a profitable opportunity:

I am not in the habit of being deterred from any experiment which I think likely to have profitable results. At my works, I do not believe there is any improvement which has come under my notice with respect to which we have not either tried it or sought out the results of other people's trials ... There is no trade in existence which has spent more money, and devoted more time and effort to get out of its difficulties than the paper trade, and I believe that there is no trade which has spent its capital for a smaller comparative profit than the paper trade during the last twenty years.[76]

Interest in finding a new source of cellulose extended beyond paper-

---

[74] *Ibid.*, pp. 342, 344, 304; Ketelbey, *Tullis Russell*, p. 107.   [75] *SC Paper*, p. 383.
[76] *Ibid.*, p. 386.

makers and free traders. In 1854 *The Times* offered a prize of £1,000 to anyone who could discover a perfect substitute for rag. To win, the inventor had to come up with a material that was practically unlimited in supply and which could produce a paper equal in quality to, and at a cost not less than 10 per cent lower than, that currently used by the newspaper. Two papermakers were appointed by the paper to test the claims of the contestants. Frederick Magney, one of these referees, reported that the exhaustive experiments on different fibres that were suggested continued for more than a year with little success, and that no prize was thus ever awarded.[77] Trade journals also were engaged in the search, frequently drawing potential raw materials to the attention of their readers. These materials were often of an exotic nature, though they also carried reports on the availability and progress of the better known materials, which were avidly followed by papermakers hoping to keep abreast of the latest developments in the trade.[78]

Attention was also shown in official circles. The Select Committee on Paper – itself the product of the public and parliamentary debate on the issue – whilst primarily established to investigate the case for the re-imposition of export duties on rags, also 'directed their especial attention to inquiring as to the possibility of applying any new fibre as a substitute for the refuse material now in use for papermaking purposes'. It concluded that, although 'great efforts' had been made a substitute remained elusive.[79] Efforts were also being made within the executive branches of government where a growing awareness of the increasing scarcity of the materials for the fabrication of paper was developing. On 13 February 1854 Wilson of the Department of Treasury wrote to the Office of Committee of the Privy Council of Trade that 'increasing scarcity has been felt of late in obtaining supplies of the raw material of paper'. He recommended:

With view to diminish the inconvenience thus felt, it has been suggested to My Lords, that Her Majesty's Consuls abroad might be instructed to obtain information, and procure samples of vegetable fibre in their respective localities, applicable to the manufacture of paper. In doing this, it would have to be borne in mind that the great essential of such an article must be its cheapness to cover the high freights now prevailing, and which, it may be anticipated, will prevail for some time.[80]

In reply, Emerson of the Office of Committee of the Privy Council of Trade on 27 May 1854 informed the Department of Treasury that the matter had been referred to the Department of Science attached to the

---

[77] *Ibid.*, p. 294. The reward was only offered for a twelve-month period.
[78] For example, see the comments of Henry Bruce, *SC Paper*, p. 304.
[79] *Ibid.*, p. 270.   [80] *Correspondence*, p. 491.

Board of Trade which would investigate the issue. In addition a report was requested from the India Board on the availability of materials in India. This report was compiled by Dr Forbes, the officer in charge of all scientific correspondence related to vegetable and plant life in India. His report, which was passed on to the Department of Treasury at the end of 1854, detailed quite extensively the many possible sources of raw material available in India.[81] However, no evidence of these materials ever being followed up by the Board of Trade or of the information being passed on to practical papermakers is available. Moreover, the nature of the report tended to view the matter from the scientific rather than the commercial aspect of papermaking; *viz.*, the question it addressed was what materials could be used to make paper, instead of whether these materials were cost-effective, as Wilson had suggested in his original letter. It was presumably for this reason that the report was never mentioned again. In any case, it was perhaps naive of Wilson to expect Forbes to address the matter of commercial viability, as such analysis could only realistically be carried out by the papermakers themselves. The information contained in such reports as well as in subsequent Board of Trade publications throughout the second half of the nineteenth century, however, did play a crucial role in suggesting alternatives to the papermaker. They are also important to us in that they exhibit the degree and depth of the efforts being made to find a solution to the shortage of raw materials.

This vigour was also expressed through the number of patents applied for especially from around the mid-century. John Evans estimated that by 1861 there had already been 'upward of one hundred patents taken out for different materials for the manufacture of paper';[82] an estimate vindicated by the data. If anything, he had underestimated the scale of the search that had gone on. In the 30 years between 1855 and 1884, 125 British applications for patents on new raw materials alone were taken out. Even allowing for the impact of the Patent Law Amendment Act of 1852, which boosted the number of patent applications made in Britain after that date, this figure still represents a marked increase over the sixty-two filed by both British and foreign applicants between 1800 and 1854. The most active period for British patenting in new raw materials for papermaking occurred between 1855 and 1864 when sixty-three applications were made, and thereafter the number of such patents steadily declined. From 1865 to 1874, thirty-eight such patents were applied for, and in the following ten years the number fell to twenty-four. This did not mean a diminution of effort in this direction after

[81] *Ibid.*, pp. 491–4.   [82] *SC Paper*, p. 292.

1865. A proper measure of the degree of innovative effort exerted to find a substitute for rag perforce has to go beyond the simple identification of a possible substitute. Processes have to be developed that make such material capable of yielding an affordable supply of cellulose. Moreover, the discovery of more efficient methods of treating rags, either to save on their use or to render previously unusable rags valuable, were also legitimate means of tackling the raw material shortage that the paper-maker was facing. Bearing this in mind, it is necessary to add those patents to do with the chemical or mechanical treatment of a known raw material prior to its arrival at the beater to the patents for new raw materials. Doing this, we find that there were in fact 601 such patents applied for by Britons between 1855 and 1884; a sum that amounted to about 60 per cent of all papermaking patents applied for by Britons in that period. The largest number of patents are found between 1865 and 1874, followed by 1855–64 and then 1875–84. The numbers for these periods respectively were 244, 202, and 155. Given that the total number of all British and foreign patents concerning papermaking in the United Kingdom in the first half of the nineteenth century did not exceed 250, it is clear that the innovative effort made by the industry between 1855 and 1874 to find a viable alternative to rag was significant and highly visible. A marked upturn in these efforts can be observed clearly from the 1850s, with the apogee being reached in the late sixties, and early seventies. We shall have reason to take a more detailed look at the pattern and composition of these patents right up to 1913 later on in this chapter, but for the meantime it should suffice to say that, as far as patent data go, there does not appear to have been any lack of vigour shown by papermakers to find and experiment with new raw materials in the second half of the nineteenth century.

A comparison

Of course not all substitutes to rag attracted the same attention from inventors and entrepreneurs, and from the 1850s efforts tended to focus on the two most likely substitutes for rag; a status these materials had attained by virtue of their ability to meet the criteria of producing a paper both at an affordable price and of a reasonable quality. The first of these new materials was esparto grass. Esparto, a generic name for the two grass species, *Stipa tenacissima* and *Lygeum Spartum*, is a tough and wiry plant with long thin leaves, that grows in clusters of anywhere between 2–10 feet in circumference and to a height of 3–4 feet. Despite the bizarre claim by one reporter in 1864 that 'esparto grass grows along the coast of Cumberland, and local papermakers are carting it to their mills in great quantity',[83] it in fact only grows in the wild in southern Spain and North Africa where it is particularly prolific at all elevations between sea-level and about 3,000 feet in the zone between 32° and 41° north latitude.[84] Of the 220,000 tons of it estimated to have been growing in the wild before its use in and cultivation for papermaking, some 25,000 to 30,000, or less than one-seventh of the total stock, was harvested for domestic consumption in the Mediterranean area. In particular, locals used the plant chiefly for basket-making, but esparto also found use as a stuffing for pillows and mattresses and as a material to make mats, casings, panniers, ropes, dishcloths, and sandals.[85] British inventors of the nineteenth century were also to find that it contained a large quantity of cellulose fibre that was relatively easily extracted and sufficient in its make-up to be used in papermaking. The first patent on esparto taken out in Britain was by Miles Berry in 1839. This was followed by two others, one in 1853 by the Parisian Jules Dehau and the other by James Murdoch in 1854. These patents, however, failed to provide a chemical treatment that could make the plant economically viable. This development awaited the arrival of Thomas Routledge's key patents of 1856 and 1860. From his mill in Eynsham, Oxfordshire, Routledge gradually discovered a successful way of rendering the material into pulp at an acceptable price. By 1858 his confidence in esparto led him to shift entirely over to the production of paper from esparto alone.[86] The process, which he pioneered and later patented, was to become the basis of all esparto preparation thereafter. The grass arrived at the mill dried and tied up in huge bales, usually sorted according to grade. These bales were broken open in a sorting hall and passed through dusters which shook off any loose dust. Following that – in what must have been an incredibly tedious job – weeds, root ends, and other impurities were picked out by hand by women. It was then stacked into 'vomiting' boilers and treated with a caustic lye, normally consisting of soda or potash and lime, and boiled at a pressure usually not exceeding 20 pounds for $2\frac{1}{2}$–3 hours. This removed the encrusting and silicious matter that coats and cements together all raw vegetable fibres and reduces the bundle of grass to a pulp. The pulp is then beaten and used to make paper in the same way as it is done with rags.[87]

The second new material to which considerable attention in the 1850s and 1860s was directed was mechanically ground wood. At this time chemically reduced wood pulp was a thing of the future. In 1840

[83] *PTR*, 1 November, 1864, p. 249.
[84] Association of Makers of Esparto Papers, *Esparto Paper* (London, 1956), pp. 1–4.
[85] Spicer, *Paper Trade*, pp. 35–6.   [86] Hills, *Papermaking*, pp. 138–42.
[87] Spicer, *Paper Trade*, pp. 15–16.

Frederich Gottlob Keller secured a German patent for a wood-grinding machine for the making of pulp for paper. This was done by pressing blocks of wood against a revolving wet grindstone, which separated the cellulose fibres in the wood from the ligninous and resinous matter. It was Heinrich Voelter, a papermaker from Saxony, however, who was to put Keller's idea on a more practical footing. After buying Keller's patent in 1846, Voelter devoted five years of his time exclusively to the practical and commercial development of the process, and together with the machine-makers I. M. Voith of Heidenheim in Württemberg produced a wood-grinding machine that was to become the standard for the German and American paper and pulp industries from the mid-century. Although virtually any soft coniferous tree can be economically converted into pulp, the types normally used by these machines were spruce, poplar, aspen, and firs, as the natural colour of their cellulose fibre was already close to white, and hence required little later bleaching. These trees were first de-barked and cut into sizes suitable for the grinding-machine. As with Keller's original model, the Voelter process called for the logs to be forced against the periphery of a rotating grindstone with the resulting chips and fibres being carried away from the grinder by a stream of water. This stream took the strips through a series of screens and filters which removed splinters and large pieces of wood but allowed cellulose fibres of a desirable length through. These fibres were then gathered together in a container, and if the paper was to be made elsewhere, these were dried and transformed into pulp boards ready for shipment to the paper-mill. In the paper-mill these boards were boiled and bleached in the same way as other pulps.[88]

A third widely known and in the 1850s probably the most tried alternative to rag was straw.[89] Since the time of Koops it had been considered by papermakers as the material with the most potential for further development. As with esparto, a number of patents for the treatment of straw were secured in the first half of the century, although none of them provided the industry with a working method that was acceptable to the majority of papermakers at the time. Indeed, it was not till John Cowley's patent of 1856 that a feasible process for the utilisation of straw was finally discovered. This process involved boiling the straw and an alkali solution, usually of quick lime or caustic potash, soda or ammonia, in an iron vessel under high pressure at $250°$ for between eight and ten hours.[90] Despite the early optimism of the papermakers for straw, its use in Britain was minimal. Apart from the very poor quality of

[88] Hills, *Papermaking*, pp. 146–8.
[89] This is certainly the impression given by the testimonies in *SC Paper*.
[90] Hills, *Papermaking*, pp. 135–6.

the paper it made, there were a number of other features of the material that acted against its profitable use by papermakers. The process was very time-consuming and profligate in its use of chemicals, which made the conversion process very costly; a difficulty compounded by the fact that the amount of cellulose that could be extracted per ton of straw was much lower than any other of the chief materials. In fact, it usually took four tons of straw to make a single ton of pulp. For this reason, the Final Report of the Select Committee on Paper of 1861 concluded that 'the greater comparative expenses of chemically reducing these [straw] raw fibres presents difficulties to their becoming a substitute for the refuse material now used'.[91] Its potential as a substitute was also undermined by its own scarcity at times, which could force the price of straw up to levels at which the papermaker would prefer to stick to his rags. In some counties, for example, farmers, bound by their leases to use their straw on the land, simply could not sell their excess straw to the paper-mills, while the storage of straw by papermakers was also not a possibility as the quality of the pulp produced from a given stock was apparently highly changeable and could deteriorate within days of cutting.[92]

Some idea of the economics of this choice can be gleaned from data available from the time. It must be said at the outset that the following cost comparisons should be taken only as an indication of the type of cost differentials being faced and not as precise calculations of the true costs. Furthermore, it should be borne in mind that there is the added difficulty that two sheets of paper made from different materials are rarely identical in appearance and quality, so that it is somewhat artificial to assume that, in what follows, we are comparing like with like. Granting this, however, it is possible to get some picture of the relative costs at the beginning of the 1860s of producing a fairly low, but popular, quality of paper – such as that used for common printing paper – from rag, wood, and esparto respectively. This can be done using the information contained in contemporary reports about experiments conducted with the various materials. The use of such data has an advantage for historical analysis over later scientific research on the materials, because it allows us to view the problem with the information available to, and from the perspective of, the entrepreneur of the time.

In January 1863 the *Paper Trade Review* published a translated report from the *Deutsche Industrie Zeitung* of the previous year, which directly compared the expenses of producing 2 cwt of a similar grade of paper from rag and from wood pulp prepared by the Voelter process.[93] Assuming that for common rags 40 per cent of its weight on average

[91] *SC Paper*, p. 270.   [92] *Ibid.*, p. 294.   [93] *PTR*, 1 January 1863, p. 42.

could be converted into pulp, this meant that some 375 lbs of rags were needed in total to produce the desired 224 lbs of pulp. In terms of wood, it was estimated that $5\frac{1}{2}$ cubic yards of fine wood was required to produce the 224 lbs. The article then reasoned that the difference in cost price between 224 lbs of wood pulp and 224 lbs of rag pulp would be equal to the difference between the price of $5\frac{1}{2}$ cubic yards of wood and the price of 375 lbs of rags plus the manufacturing costs of wood pulp over and above those associated with the preparation of the rags. The article priced common rags at 9s 2d per cwt and wood at 1s 7d per cubic yard. This seems to be a reasonable price for rag, as rag suitable for the manufacture of common printing paper in Britain could be had for between 9s and 12s per cwt.[94] In fact, the price given in the article would be at the lower end of rag prices, so if anything, it would bias the findings in favour of rag in the following comparisons. We know that the conversion rate for wood into pulp ranges from between 55 and 70 per cent of its weight. If we assume an even lower bound estimate of 50 per cent as the actual conversion rate – an assumption that would tend to overstate the cost of wood – then we can expect that for each 2 cwt of paper produced, we would need 4 cwt of wood. Alexander Cowan and Sons, a Scottish papermaker experimenting with wood in the 1860s, claimed to be able to get fir for £1 15s per ton and wood that would make the 'best pulp' for £2 12s 6d per ton.[95] This suggests that the cost of 4 cwt of wood to British paper producers ranged somewhere between 7s and 10s 6d. This would put the article's price of 8s 8d for approximately 4 cwt well within the feasible range. We will, however, adopt 10s 6d per 4 cwt (i.e. 2s $7\frac{1}{2}$d per cwt or 1s 10d per cubic yard) as our price for wood, since this would also tend to bias the calculations in favour of rag. The modified results of the experiment are as follows.

(1)	The cost of 375 lbs of rag at 9s 2d per cwt	£1 10s 7d
(2)	5.5 cubic yards (or 4 cwt) of wood at 1s 10d per cubic yard (or 2s $7\frac{1}{2}$d per cwt)	10s 6d
(3)	The cost of preparing mechanical wood pulp minus the cost of preparing rag pulp	4s 2d
(4)	The difference in favour of wood pulp $\{1-(2+3)\}$ per 224 lbs of pulp	15s 11d

[94] *PTR*, 1 September 1863, p. 177.
[95] Cowans Collection, GD 311/1/9, Logbook 20 August 1867.

The difference in the cost of producing 224 lbs of paper from rag and the same amount from wood pulp was thus just over 15s. Per ton of paper that was the equivalent of £7 19s 2d: quite a significant margin. We can also use these calculations as the basis for a similar comparison of esparto with rag, as well as esparto with wood.

Jules Barse's investigation of esparto, or the alfa fibre as it was also called, published in 1864 proclaimed the viability of this fibre for papermaking. Crucial to his calculation was the finding that 73.5 per cent of the weight of the esparto could be turned into paper stock.[96] This was an optimistic assertion and it is highly unlikely, given the inchoateness of the techniques for the extraction of the cellulose, that practical papermakers could get anything like that percentage of cellulose from the esparto they used. A 50 per cent conversion rate was probably quite common, but we shall assume 66 per cent as an average which the papermaker could expect. The highness of this percentage is desirable as it will also bias the comparison of esparto and wood pulp in favour of esparto. Given this conversion rate, we would thus require 340 lbs of esparto grass to produce 224 lbs of pulp. The price of esparto in Britain in the early 1860s fluctuated between £5 and £7 per ton.[97] Taking £6 as the average of these prices, we can estimate that this quantity of esparto would cost about 18s 5d.

The major difference in the manufacture of paper from esparto and from rag was the former's use of more chemicals: namely caustic soda. In other respects – labour, equipment, and power – there was little cost difference between the materials. This is implicit in the articles already referred to and was generally believed at the time.[98] This relatively greater utilisation of caustic soda in papermaking with esparto needs to be taken into consideration in our calculations. Lower grades of rag paper needed about 1 per cent of its weight in caustic soda, so that to make 224 lbs of such paper from rags required approximately 4 lbs of soda.[99] This can be contrasted with esparto which used as much as 13 per cent of its weight and possibly even more with certain batches.[100] This meant that around 44 lbs of caustic soda would be needed to turn the 340 lbs of esparto into a comparable quantity and quality of pulp as made with common rags. Subtracting rags' requirements of caustic soda from esparto's, we find that esparto paper needs about 40 lbs more soda than rags to make 2 cwt of paper, or alternatively 3.57 cwt more per ton

[96] PTR, 1 March 1864, p. 73.
[97] Spicer, Paper Trade, p. 33; PTR, 1 July 1863, p. 146.
[98] W. S. Shears, William Nash of St. Paul's Cray Papermakers (Leeds, 1967), p. 27.
[99] Spicer, Paper Trade, p. 12.
[100] Association of Makers of Esparto Paper, Esparto, p. 4.

of paper produced. Given that the price of caustic soda in Britain in the 1860s stood at around £7 per ton, this difference would amount to an extra outlay of 2s 6d on soda for esparto users for each 2 cwt made.[101] This expenditure on soda could be reduced if the caustic liquor which was expelled from the digester after the boiling process could be recovered and re-used. Routledge took out two patents for processes with this intent in 1865 and 1866, but neither of these proved very effective until the Porian evaporator in 1877 and the multiple-effect apparatus in 1886 had been invented.[102] In the early 1860s, however, none of these devices were available, so the papermaker continued to dispose of his used liquor in a neighbouring river and purchased new soda each time he decided to make a fresh run of esparto paper. Putting together this information, we come up with the following comparative cost figures.

(i)	The cost of 375 lbs of rags at 9s 2d per cwt	£1	10s	7d
(ii)	The cost of 340 lbs of esparto at £6 per ton		18s	5d
(iii)	The cost of 40 lbs extra caustic soda required for esparto at £7 per ton		2s	6d
(iv)	The cost of 5.5 cubic yards (or 4 cwt) of wood at 1s 10d per cubic yard (or 2s 7½d per cwt)		10s	6d
(v)	The cost of preparing mechanical wood pulp minus the cost of preparing rag/esparto pulp {not including (iii) above}		4s	2d
(vi)	The cost difference in favour of wood over rags per 224 lbs of pulp {(i) − [(iv)+(v)]}		15s	11d
(vii)	The cost difference in favour of esparto over rags per 224 lbs of pulp {(i) − [(ii)+(iii)]}		9s	8d
(viii)	The cost difference in favour of wood over esparto per 224 lbs of pulp {[(ii)+(iii)] − [(iv)+(v)]}		6s	3d

Clearly esparto paper was less expensive to produce than rag paper (by 9s 8d or £4 16s 8d per ton of paper). The interesting feature about these calculations is that not only is paper using mechanically ground wood pulp cheaper than paper made from rag, it is also cheaper than paper manufactured from esparto (by 6s 3d or £3 2s 6d per ton of

[101] Spicer, *Paper Trade*, p. 78; Hills, *Papermaking*, p. 139.
[102] Hills, *Papermaking*, pp. 141–2.

paper). Before we can read anything into these findings, two matters still have to be considered. Firstly, some allowance must be made for internal transport charges within Britain itself. For mills not located near the big ports, the prices of raw materials cited above may not reflect the additional costs of getting that material further inland. This would only be a problem if there were differences in the internal cost of transportation between the materials. Yet even if this were the case, it is doubtful whether this factor alone could materially alter the basic tenor of the above calculations. For example, assuming that each ton of wood paper needs two of wood, for internal costs to erode wood's cost advantages over esparto totally, wood would have to cost £1 11s 3d more per ton to transport to the mill from the port than esparto. This seems highly unlikely. In November 1867, Alexander Cowan and Sons had 1.5 cwt of wood brought from Glasgow to its mill in Penicuik – a distance of about 46 miles – at the rate of 5s per ton, which is way below the figure quoted.[103] Moreover, this was the full cost of the transportation, not just the excess of wood's over esparto's cost of transportation. Since most mills did not lie much further from a major port than 50 miles it is hard to imagine internal transportation costs playing an important role.[104]

Secondly, according to neoclassical theory, what papermakers were really interested in were the *total* costs of the alternative material relative to the *variable* costs of using rag.[105] The above makes no allowances for the cost of any new machinery needed to use the new material. These, however, at least viewed *ex ante* from the perspective of the papermaker would have been negligible. It was widely believed at the time that esparto could be used with existing plant. With respect to esparto, Charles Cowan noted in the jury report of the 1861 Exhibition: 'one satisfactory feature in Mr. Routledge's process is the fact that no material alterations in existing machinery or appliances are required ...'[106] This was no chance remark; the *Paper Trade Review* repeated it on other occasions such as when it reported that, 'no material alterations in the machinery or apparatus is required for working esparto ... The successful working of this fibre depends mainly on the careful and proper adjustment of the quantity and strength of the chemicals employed.'[107] This is fundamentally true. Esparto paper can be made

---

[103] Cowans Collection, GD 311/1/9, Logbook 12 November 1867. This result would hold even for the figures given for a later period in W. K. Lawson, *British Railways: A Financial and Commercial Survey* (London, 1913), pp. 207–8.

[104] For example, the two other great papermaking cities of Britain, Manchester and Maidstone, lay within this distance from ports. Manchester lies about 35 miles from Liverpool and Maidstone around 38 miles from London.

[105] Salter, *Productivity*, pp. 48–65.   [106] *PTR*, 1 July 1863, p. 145.

[107] *PTR*, 1 September 1863, p. 177.

with the same equipment as rag paper. This, of course, is not to say that in time it was not found prudent to alter the machinery to better meet the properties of the esparto process, but this was only part of the natural learning process and not the development of a totally independent technology. The rotating spherical boilers which had been used for rags, for example, were not ideal for boiling esparto, since they lacked the mechanical strength to withstand the pressure required for the boiling process. Papermakers and machinists experimented with alterations to their existing equipment, and eventually with the introduction of the fixed Sinclair Esparto Boiler in the 1870s this problem was solved. Rag willows and dusters also underwent some adaptions for their use with esparto.[108] None of this, however, would have gone into the considerations of the papermakers of the late 1850s and early 1860s, who experimented with esparto on their existing equipment and did not envisage a move to esparto entailing more than a very minimal expenditure on new equipment.

Mechanically ground wood pulp, however, did need new machinery. Unfortunately, as few if any were actually purchased in Britain at this date, no price data for such machinery are available. What we can do, however, given a few assumptions, is to estimate how much a papermaker would have been prepared to spend on such machinery and still be no worse off from utilising wood instead of esparto. That amount would be equal to the net present value (NPV) of the expected return of shifting to wood from esparto. This can be calculated with the formula

$$NPV = \sum_{t=0}^{\infty} R_t/(1 + r)_t - K$$

where R is the annual gain from using wood instead of esparto, r the discount rate, K the cost of the machinery, and t time. As long as the NPV of the project is zero or greater, the investment in new machinery should go ahead. By setting the NPV to zero – the borderline case – this allows us to work out the maximum amount the papermaker would be prepared to pay for the wood-grinding equipment. What needs to be calculated then is the current revenue flow of the investment. By using extreme values for R, r, and t, we can reduce K and heavily bias the calculation against the use of wood. Our analysis has suggested that the savings in cost accruing to the producer, who used wood instead of esparto, came to around £3 2s 6d per ton of paper produced. If we assume that the wood-using mill only produces 250 tons of paper per

---

[108] N. Watson, *The Last Mill on the Esk: 150 Years of Papermaking* (Edinburgh, 1987), p. 27.

annum (an annual saving of £781 5s), that the rate of discount is 5 per cent, and that the machinery is expected to last only seven years after its purchase, then it is worthwhile for papermakers to introduce wood-grinding machinery up to the value of £5,301 1s 5d. Given that in 1863, the buildings, grounds, plant, and machinery of Moffat Paper Mill, a single machine-mill near Airdrie, went on sale for £6,000, then £5,300, which is nearly 90 per cent of the entire cost of a such a standard running mill and more than the largest Fourdrinier of the time cost, would seem a ridiculously high price to pay for wood-grinding equipment.[109] The actual price of wood-grinding machinery almost certainly must have been less than that, so that the cost of such machinery was unlikely to have significantly eroded the cost advantage identified above. It should be reiterated that all of the three assumptions made above are quite heavily weighted against the use of wood pulp. In 1865 average output per mill, inclusive of hand-mills, was around 300 tons per annum, interest rates were below 5 per cent, and grinding machinery had a life expectancy much greater than a mere seven years.[110]

Fixed costs were thus not an important aspect of the choice as it presented itself to the papermaker at the mid-century. We are left then with our original findings that both esparto and wood paper were almost certainly cheaper to produce than rag paper, though wood paper by a greater margin than esparto paper. Looked at in this way, a papermaker with this information should have opted for wood production, or at least begin to mix wood with rag. That they did not calls for some explanation.

That explanation lies in the definition of the problem tackled by the papermaker. If we were talking about perfect substitutes and the only objective of the papermaker was to find a replacement for rag, then the choice of esparto would be peculiar. However, if what papermakers were actually looking for was a material, that, although perhaps less than ideal, could meet certain standards and could thus act as a temporary and partial replacement for rag whenever there was a shortfall in its supply, then the decision may be more comprehensible. Obviously, had a perfect substitute in every way to rag appeared in 1860, there is no reason to believe that papermakers would not have clamoured to get their hands on it. Sadly, this was not the case. 'There is nothing equal to rags', announced Thomas Chalmers in 1861; a sentiment backed up by virtually every other papermaker.[111] The comments of John Evans, the

[109] *PTR*, 1 October 1863, p. 200.
[110] Spicer, *Paper Trade*, appendices V and IX; E. H. Phelps Brown and S. A. Ozga, 'Economic growth and the price level', *Economic Journal* 65 (1955), 2.
[111] *SC Paper*, p. 300. Also see Evans, Chater, Bruce, and other's comments on pp. 293–4, 296, 305, 341, and *passim*.

president of the papermakers' national association, at a meeting of the trade in London in 1864 also implies very strongly that dominance of rags was still seen as certain. In response to a question about new materials, he replied 'that the increase of new fibres was more injurious than beneficial. For instance, if you have an export duty as now of £5 upon rags costing £20, you have the present percentage; but if by the increase of new fibres, rags go down to £15, and the duty is continued at £5, you have a percentage of thirty-three to contend against.'[112] Whether or not this mattered to papermakers, it does at least illustrate that a key member of the trade, and one certainly not shy of change, could not conceive of rags ever being replaced, and that he saw that the primarily purpose of new materials was to act as leverage on the price of rag. As Evans continued, 'it is only when rags are at certain price, that the article esparto – a sort of grass – can be used'.[113]

The failure of materials other than rag to elicit the respect of papermakers was chiefly due to their inability – at least at this time – to replicate the quality and appearance of rag paper. Referring explicitly to straw, though it holds *a fortiori* for mechanically ground wood paper, Saunders, for example, could see little scope for the use of such new materials, as with them, 'you would not get the same article either for use or in appearance'.[114]

In terms of quality, although esparto paper in the 1860s was still patently inferior to good rag paper, it clearly outstripped paper made by the Voelter process. This was because the fibres of mechanically ground wood pulp were short, inflexible, and without the improvement later experimentation was eventually to provide, could not bond together very well because of the resinous and gummy material and lignin of the original wood. Esparto fibres, although also short, were more cohesive, mixed well with other pulps, and could produce a paper that was strong, smooth, and absorbent. Once worked out esparto thus could be used to make a paper approaching the quality of rag paper.[115] This was beginning to be appreciated at the time. A highly influential paper-making manual of the 1870s saw esparto, if not in glowing terms, as being of some value to the papermaker:

Esparto does not possess the same strength or tenacity as the fibres produced from flax, hemp, cotton or even jute; its strength, indeed, is fictitious, due to the

---

[112] *PTR*, 1 July 1864, p. 173.   [113] *Ibid.*, p. 179.
[114] *SC Paper*, p. 303. American manufacturers of the time also showed the same initial resistance to non-rag fibres. See chapter 7.
[115] 'the mechanical pulp then available [1876] was unsuited to fine papermaking ... the relative costs of esparto and wood have been important in determining the proportions of each used, although quality has always been first consideration'. Weatherill, *One Hundred Years*, p. 20. Also see, Hills, *Papermaking*, pp. 138, 147–8.

126 Productivity and performance

gum, resin, and gluten combined with the true fibre; its value as a paper-making material being dependent upon its comparatively low price, the facility with which it can be blended with other material, and the simplicity of the process of treatment.[116]

As a result of these qualities, it claimed esparto had become a mainstay of the British paper industry by 1876. The manual's assessment of wood pulp was less optimistic:

Lastly may be mentioned wood, which has created much interest in the trade of late, and has attracted to itself a considerable amount of experimental attention; but so far as regards the English trade, results have not been sufficiently satisfactory to lead to anything more than its exceptional adoption.[117]

Given the experience with and state of knowledge about the use of wood for papermaking available in Britain at this time, wood pulp just could not meet the needs of the industry.

The problem as it was perceived by the papermaker of the mid-nineteenth century was not necessarily to discover a perfect substitute for rag, rather to find a material that could ease the pressure on the supply of rag. By introducing a small percentage of such a material, a papermaker could lower his costs somewhat without necessarily jeopardising the integrity of the paper too much.[118] This was the primary concern and objective of most of those experimenting with new materials both in Britain and elsewhere in this period. In the United States, for example, their papermakers too at this time still only looked to blend relatively small quantities of the new raw materials with rags.[119] Back in Britain, George Chater expressed the prevailing view succinctly when he told the Select Committee on Paper in 1861, 'I think no substitute can be found for rags. It strikes me as an impossibility.' He went on to add, however, that, 'something to *assist* [my italics] might be found ...'[120] This was also the view that appeared in the final report of the Select Committee. After explaining that the committee had seen little evidence of success in finding a substitute for rag, the report could still conclude that it saw 'no reason to doubt that straw, and other fibrous substances, may form a *supplementary* [my italics] part of the material for papermaking ...'.[121] Before the advent of chemically reduced wood pulp the material most likely to fill this supplementary role was esparto.

[116] Paper Mills Directory, *The Art of Papermaking: a Guide to the Theory and Practice of the Manufacture of Paper, being a Compilation from the best known French, German, and American Writers* (London, 1876), p. 100.
[117] *Ibid.*, p. 6.   [118] *Ibid.*, p. 345.
[119] For example, L. Calder, *The First 100 years: Perkins-Goodwin Company* (New York, 1946), p. 19.
[120] *SC Paper*, p. 341.   [121] *Ibid.*, p. 270.

Esparto could be used with and in place of rags because it showed itself able to yield a somewhat comparable product to that produced by rag at a cheaper price. The mechanical wood pulp familiar to British producers in the 1860s, however, could only claim to meet the latter of these two criteria.

A case study

The best way to examine how the British industry tackled the process of finding and developing new raw materials is to look at it as it occurred at the firm level. A firm for which reasonable records exist for this period is Alexander Cowan and Sons Limited of Penicuik near Edinburgh. Operating from its Valleyfield Mill from 1709, the firm had a long tradition of producing a variety of reputable paper, ranging from tub-size writings[122] right through to printing and newsprint. Its paper was chosen by Sir Walter Scott for the first editions of his novels and found markets as far away as Australia.[123]

Cowans was also one of the first companies to become very interested in the use of esparto grass. On 12 July 1861 it entered into a ten-year agreement with Thomas Routledge that made him the firm's sole supplier of esparto.[124] This half stuff usually came from Ford Mill, near South Hylton in Sunderland, which had been set up by John Dickinson and Thomas Routledge just a year earlier.[125] The agreement guaranteed that the firm would get its prepared esparto for 16s on the original weight of the esparto. It is not clear how the firm first became aware of esparto or Routledge's process, but at first it evidently was pleased with the arrangement and the pulp it received: 'it is well prepared and dried and clean and well bleached and obviates doubtlessly some of our pollution'.[126] But it is equally evident that esparto was intended as a stopgap measure, and not as a material that would supersede rag. On 30 August 1864 it was noted in the firm's logbook:

ESPARTO – we must continue it for a short time as a sufficiency of Prints has not been got – but we are resolved at an early period to cease its use. Although we are and have been for some time treating it very successfully.

[122] A type of quality writing paper where a size is applied after the paper web has been dried by immersing the paper in a mixture of alum and warm gelatine. This reduces the rate at which the paper absorbs water. Hills, *Papermaking*, p. 223.
[123] *PTR*, 12 July 1912, pp. 61–2; Spicer, *Paper Trade*, p. 215; notes on the firm's history listed in Cowans Collection, GD 311/7/66.
[124] Cowans Collection, GD 311/1/8, Logbook 30 October 1866 and 5 March 1867.
[125] Evans, *Endless Web*, p. 111.
[126] Cowans Collection, GD/311/1/8, Logbook 30 October 1866. Similar entries can be found on 30 August 1864 and 3 January 1865.

Still, the firm was satisfied enough with the progress of esparto to have made plans by August 1866 to set up a laboratory to find ways of extracting cellulose from esparto without the noxious by-products of existing processes that currently polluted the local rivers. This apparently had some success, resulting in a patent being taken out.[127] These experiments were also very important because they gave the firm the necessary expertise to produce its own paper stock and thus break its dependence on Routledge for esparto pulp. For this purpose the firm in 1867 bought a property at Musselburgh for £1,400, where it built a 20 ton a day esparto-mill.[128]

An important stimulant to improvements in the processes used to make esparto paper was the increasing agitation against the pollution of the rivers it caused. This agitation reached its peak in 1867 when the Duke of Buccleuch, Lord Melville, and others successfully took the seven papermakers of the North Esk to court over the damage such discharges had done to the river. In 1869 the parties in the lawsuit agreed to place the matter in the hands of Professor Perry of Glasgow with view to the discovery and application of remedial measures. Following Perry's death at the end of that year this role was taken over by W. Arnot. All costs were to be borne by the papermakers and by September 1869 these already amounted to £837.[129] Perry not only visited the mills to ensure compliance to the guidelines, but also to make useful suggestions to the papermakers. For example, on one such trip in May 1869, 'Dr Perry conducted an interesting series of experiments upon the precipitation of rag and esparto washings by means of Alumina and milk of lime. These were wonderfully successful and the Dr further showed us the effect of the addition of a fluid which resulted in the almost immediate suppression of froth.'[130] This advice permitted the firm to reduce costs by affording a greater recovery of soda. Earlier the firm also seems to have hired its own chemist between 19 June and 22 October 1866 who also conducted experiments on the reduction and recovery of the spent caustic lye.[131] This resulted in continued improvements in the use of esparto not only with respect to pollution, but also the quality of paper it could produce. As the entry of 16 July 1867 could report about esparto use, 'we used last week 26 tons, and from the better ability now to treat it, it is preferable to raw rags'.[132] A cost comparison of rag and esparto at the time confirmed the relative cheapness of the

[127] *Ibid.*, 28 August 1866 and 2 October 1866.   [128] *Ibid.*, 5 March 1867.
[129] Cowans Collection, 311/1/9, Logbook 19 October 1869.
[130] *Ibid.*, 25 May 1869.
[131] Cowans Collection, Wahal's handwritten memorandum, GD 311/7/31.
[132] Cowans Collection, GD 311/1/9, Logbook 16 July 1867.

latter, with a pound of rag paper costing 4s and the equivalent weight of esparto paper only 2s 2d. 'It will thus be seen that esparto is much the more profitable material of the two, and when we work down our stock of rags, and get the number of cutters gradually reduced, we shall probably use a larger proportion of esparto.'[133]

This rise in the importance of esparto had much to do with the improvements the firm had gleaned from its own experience with the material, as well as from the information it could get from others familiar with it. In less than a decade Cowans had transformed esparto from an experimental to a proven papermaking material. This shakedown period was absolutely essential for the development of the material and took place simultaneously in many mills across the country. Thomas Wrigley of Bridge Hall Mills, for example, experimented on numerous occasions with esparto, as did the firm C. E. and J. G. Potter of Darwen, Lancashire, before they too had success with esparto.[134] It was, of course, a period not without its problems; problems which at times must have threatened the project's very continuation and which called for the same entrepreneurial drive that had initiated the project to see it through to its fruition. Without this period of learning sufficient information for rational and realistic calculations could never have been collated. One rider must be added. Esparto, despite its growing importance as a raw material within the firm, did not completely lose its label as a mere supplement for rag, as most of the increase in its utilisation in these years was due to its greater use in the pulp of 'low priced papers'.[135] Esparto was nevertheless steadily gaining ground in this firm as it was in many others.

Despite the improvements made in esparto paper, the firm's experimentation with other materials did not stop. Mounting interest was especially exhibited in wood pulp. In 1863 Charles Cowan received a visit from Voelter, the inventor and patentee of the machinery for pulping wood. Together they made a tour of the Highlands for the purpose of 'shewing Mr. Voelter how abundantly common fir could be found'. Voelter, however, recommended that a white wood be sought out, as this alone he thought was capable of yielding a pulp which would serve as a substitute for rag. He advised strongly against Scots fir as he

---

[133] Ibid.
[134] M. Tillmans, Bridge Hall Mills: Three Centuries of Paper and Cellulose Film Manufacture (Tilsbury, 1978), p. 50; SC Paper, p. 342. A similar shakedown period occurred with the adoption of kraft pulp in the southern states of the United States in the interwar period. See A. J. Cohen, 'Factor substitution and induced innovation in North American kraft pulping: 1914–1940', Explorations in Economic History 24 (1987), 197–217.
[135] Cowans Collection, GD 311/1/9, Logbook 18 January 1869.

thought it too resinous and stubborn in bleaching. Cowans do not seem to have accepted Voelter's assessment, although it took them another four years before the firm actually got around to testing Scots fir; a test that was motivated by 'the desire to diminish the pollution of the stream by the use of materials requiring no chemical preparation'.[136] The trial of 5.5 cwt of dry pulp from Scots fir produced by Voelter's machinery at nearby Morepth mill proved encouraging, the wood having blended with the rags and esparto so thoroughly as to be perfectly similar. Enquiries as to the cost of supplying the firm with the wood pulp quickly ensued. When Mr. Birrell visited Weir in June at his mill at Morepth he was informed that the cost of his best pulp stood at £9 12s 6d per ton of pulp. This was considered rather steep, but negotiations between the two parties continued. In a letter from Weir of 15 August 1867 the firm was further advised that the cost for preparing 53 cwt per diem of dry wood pulp with four Voelter machines was actually £8 16s 9d. The reduction in the price Weir was offering to Cowans had its desired effect, for it was decided, 'we are to have another trial of the Morepth halfstuff and resolve to make further enquiry as to the probable extent of the supply of wood pulp fit for our use, to be obtained from Continental mills'.[137] The second trial, however, was unsuccessful, and the firm decided not to try any more of the pulp. Two cwt of Danish dry wood pulp, bought from Leith at 11s per cwt, was also tried in September of that year, but as in the case of Weir's pulp the dinginess of the colour and the presence of a large number of large knots that had not been thoroughly beaten out disfigured the paper.[138] Around this time, Cowans was approached by D. A. Fyfe of Glammis, Forfarshire, who offered to supply bleached wood pulp made from sawdust at £6 per ton. The firm agreed to give Fyfe's process a trial, but limited expenditure on the trial to £50. Fyfe produced a bowlful of his sawdust pulp, which looked 'good, clean, and white'.[139] However, when it came to be worked up into paper the following week, the result was once again a disappointment: 'the paper turned out full of clear specks and very objectionable ... There was great froth in the engine and probably a much larger waste than Mr. Fyfe calculated. Mr. Fyfe has not been here since he wishes to put it through the stoner again before washing in the engine.'[140]

Fyfe's sawdust pulp did not get much further, but another of Fyfe's processes for treating shavings seemed to show more promise and the firm agreed to go shares in a limited company with McLean and Hope and J. Brown and Co. A pulping mill was established at Kilbagie with

[136] *Ibid.*, 20 August 1867.   [137] *Ibid.*   [138] *Ibid.*, 1 October 1867.
[139] *Ibid.*, 12 November 1867.   [140] *Ibid.*, 19 November 1867.

Mr. Fyfe as manager. Following their disappointments with mechanically ground pulp, emphasis in this new facility was put on the development of chemical processes of reducing wood. By March 1868, Fyfe, however, had already been dismissed for making 'unreasonable' claims and his shavings experiments which had already cost £160 were terminated. R. C. Menzies took over the experimental mill, working on a nitric treatment for wood and sawdust that would produce a good paper at an affordable cost.[141]

In the meantime new chemical processes were being developed elsewhere and the firm was always prepared to give promising techniques a trial. In 1869 Mr. Birrell reported back to a meeting of the firm's owners about a process owned by the Gloucestershire Paper Co., which he had witnessed on a trip to Cardiff. This was essentially the soda process for making wood pulp invented by Hugh Burgess and Charles Watt in England in 1851. Pine imported from the Baltic was broken into chips and boiled at a temperature of 400°F in caustic soda at the rate of 36 lbs per cwt. After boiling the wood pulp was washed and bleached in the same way as esparto, and apparently resembled that pulp in terms of colour and length of fibre. Birrell described it as 'beautiful', though because of the quantity of alkali it used it was more expensive than esparto pulp. Unperturbed by this, Birrell replicated the Cardiff experiment at Kilbagie, but with similar results: a ton of the wood pulp costing 22s 10d compared with 19s 10d for esparto pulp.[142] Despite the quality of this wood pulp, its costs relative to esparto were considered too great to adopt the process.

The experiments begun by R. C. Menzies at Kilbagie in 1868 continued and by 1872 he was believed to have made sufficient progress to warrant taking out a patent on his work. However, persistent teething problems prevented its commercial use. In July 1872, a trial of the process with Menzies present was undertaken, but proved disastrous because the chlorine was admitted so rapidly that the wood was all charred.[143] Menzies left, promising to prepare another batch which would be more successful. From his failure to be mentioned again in the company's records, one can only assume that Menzies was unable to fulfil this promise.

This also marked the end of the firm's independent efforts to develop a process for chemically reducing wood. Thereafter, attention – and there seemed quite little of it for the rest of the 1870s – seemed to have

---

[141] *Ibid.*, 12 and 18 January 1869; 28 January, 24 March 1868.
[142] *Ibid.*, 20 and 27 July 1869.
[143] *Ibid.*, 13 February, 9 July 1872. For other failed attempts by Menzies, see the entries for 19 March, 25 June 1872.

been devoted solely to testing the methods patented and the pulp prepared by others. This involved making trips to Sweden and elsewhere to see the preparation of the pulp firsthand. One Swedish company in 1873 offered Cowans a partnership in a new pulp-mill there, but the firm declined the offer after being dissatisfied with the samples sent.[144] The appearance of various commercially viable methods for chemical extraction, at the beginning of the 1880s led to a renewal of efforts in this direction by the firm.

There were two new methods available from that time onwards: the sulphate and the sulphite processes. The sulphate process was very similar to the soda process already familiar to papermakers. It was invented by Carl F. Dahl in 1884. With the sulphate process, wood shavings are boiled in a three to one solution of sodium sulphate, then washed and dried out. Its advantage over the older soda process stemmed from its heavy use of sodium sulphate, instead of caustic soda, which was cheaper and showed itself more readily recoverable than the soda. This made the sulphate process much more economical to papermakers than the soda process.

By contrast, the sulphite process involved treating wood shavings with a solution of bisulphite of lime, or magnesia, under a pressure of about seven atmospheres for a period of from eight hours to three days depending on the exact method applied and the type of pulp desired. The initial work on the sulphite process was undertaken by the American chemist, B. G. Tilghman. His patent of 1867 in many ways anticipated all subsequent patents in this field, yet he could not make his discovery into a commercial success due to engineering difficulties with the boilers he designed. Carl D. Ekman, a Swedish chemist, continued Tilghman's work in the 1870s and by about 1879 had made the sulphite process viable enough to have it introduced into Ilford Mills near London. Five years later the Ekman Pulp and Paper Company oversaw the erection of large mills at Northfleet using Ekman's version of the sulphite process. Other versions were devised by Partington in the UK, Francke in Sweden, and Mitscherlich in Germany.[145]

The sulphite process in particular attracted the interest of Cowans. On several occasions Birrell visited Ilford Mills in Essex to see for himself 'the usefulness of wood pulp prepared with Ekman's process' and found that, 'the paper made there [was] wonderfully good, and as well sized as if gelatine were used'.[146] He was also invited along with

---

[144] *Ibid.*, 7 April 1873. For other experiments with foreign pulp, see remarks on 8 December 1873.
[145] Hills, *Papermaking*, pp. 150–3.
[146] Cowans Collection, GD 311/1/10, Logbook 25 September 1882.

R. C. M. Spicer and J. Evans on a trip to Sweden to see the pulp-mills at Bergwerk and to witness in person the Francke process at Horndal. On his return Birrell accompanied Spicer back to Godalming, Surrey, to examine the Ekman process as it was operated there.[147] The logbook entry of 30 October 1882 makes it clear that by this date Cowans had come round to believe that, from the information available, wood pulp was now a viable and cost-effective material and that the firm ought to use it. The question remained as to which of the competing patents to opt for. It essentially came down to a choice between the patents of Ekman, Francke, and Newton (holder of the Tilghman and Mitscherlich patents). There was some debate whether these patents overlapped and legal opinion was sought to advise on which process had precedent and was closest to the method intended to be introduced at the firm's mill in Kilbagie.[148] In obtaining legal advice, the firm was not being unduly cautious. Because of the close resemblance of the different versions of the sulphite process, many court battles over the priority of the respective patents were contested between the owners of rival patents.[149] On the basis of this legal advice, Cowans took out a license on the Newton patents. This decision illustrates that by this stage the various options available to the papermaker had already been sufficiently well spelled out to permit the decision to be made on the basis of existing information alone. By the beginning of 1883 the licence was already in operation, and wood pulp had become an important raw material for the company.[150]

Mapping the transition

Cowans' story is compatible with the aggregate experience of the industry. Figure 4.1 attempts to capture the shifting relative importance of the various raw materials between 1861 and 1913.[151] This is done by looking at the share of each of the materials out of the total tonnage of raw materials used in Britain. Assuming that for the whole industry each material had on average a fairly similar conversion rate of material weight into pulp, then these figures also give a broad indication of the

---

[147] *Ibid.* [148] *Ibid.*, 30 October 1882. [149] Spicer, *Paper Trade*, p. 23.
[150] Cowans Collection, GD 311/1/10, Logbook 1 January 1883.
[151] Figure 4.1 is compiled from information in the *Annual Statement of the Trade and Navigation of the United Kingdom*. Lewis, *Numerical Approach*, p. 30 makes similar estimates based on import figures. However, as his figures do not make allowances for domestic rags, which made up the vast majority of that material used in Britain in the nineteenth century, they significantly underestimate rag's continued importance up to 1890, and consequently, also overstate the rapidity of the uptake of both esparto and wood.

percentage of paper made from each material. Even if we drop this assumption, Figure 4.1 still highlights the trends in the usage of the various materials,which is what we are chiefly interested in. The quantity of wood pulp (chemical and mechanical) and esparto used comes from import figures, since very little of these materials did or even could have come from the United Kingdom itself. One problem that needs to be noted is that between 1870 and 1886, when wood usage was still in its infancy in this country, vegetable matter other than wood and esparto was also contained in the figures for wood. This, however, ought to have very little effect on the overall total calculated, except to inflate the importance of wood slightly in this period. It is hard to believe that this fact would substantially alter the picture given by Figure 4.1. Rag usage is much harder to gauge, as rags for papermaking were obtained both at home and from overseas. Luckily there appears to have been some relationship between total rag usage and its importation. Figure 4.1 thus employs the assumption, based on contemporary belief, that the United Kingdom's importation of rags consistently represented up to 20 per cent of the country's entire requirements for rags; this gives us some idea of the scale and importance of rag use in this period.[152] This is, of course, not a precise figure. Coleman does suggest a less important role for imported rags than we do.[153] However, as we know rag did steadily lose its share, and as we are, after all, more interested in the progress of the other materials in these later periods, any possible underestimation of rag usage implicit in the calculations is unlikely to be sufficient to alter significantly the nature of the picture given by Figure 4.1.

Three distinct phases are visible in Figure 4.1. The first occurred in the period between 1861 and 1877 in which the rise of esparto took place. This was a fairly quick rise (Table 4.1). From negligible levels in 1861, within a mere three years its use had already jumped to 43,000 tons: around a third of that of rags. This steady rise continued, except for a few years between 1868 and 1872 when it hovered around the 50 per cent mark, until the use of esparto reached its zenith in 1877 at 63 per cent of the tonnage of all raw materials used in papermaking. Its growth in importance was of course solely at the expense of rag, which even as late as 1877, when wood was just about to make its presence felt, still represented 30.5 per cent of the total tonnage. The rise of esparto in the 1860s can be attributed largely to the fact that at the time it was the only material that was acceptable to papermakers. Further experimentation and learning by doing and by using strengthened the position of esparto vis-à-vis other materials in Britain right up to the mid-1870s,

---

[152] SC Paper, p. 283.   [153] Coleman, British Paper Industry, p. 214.

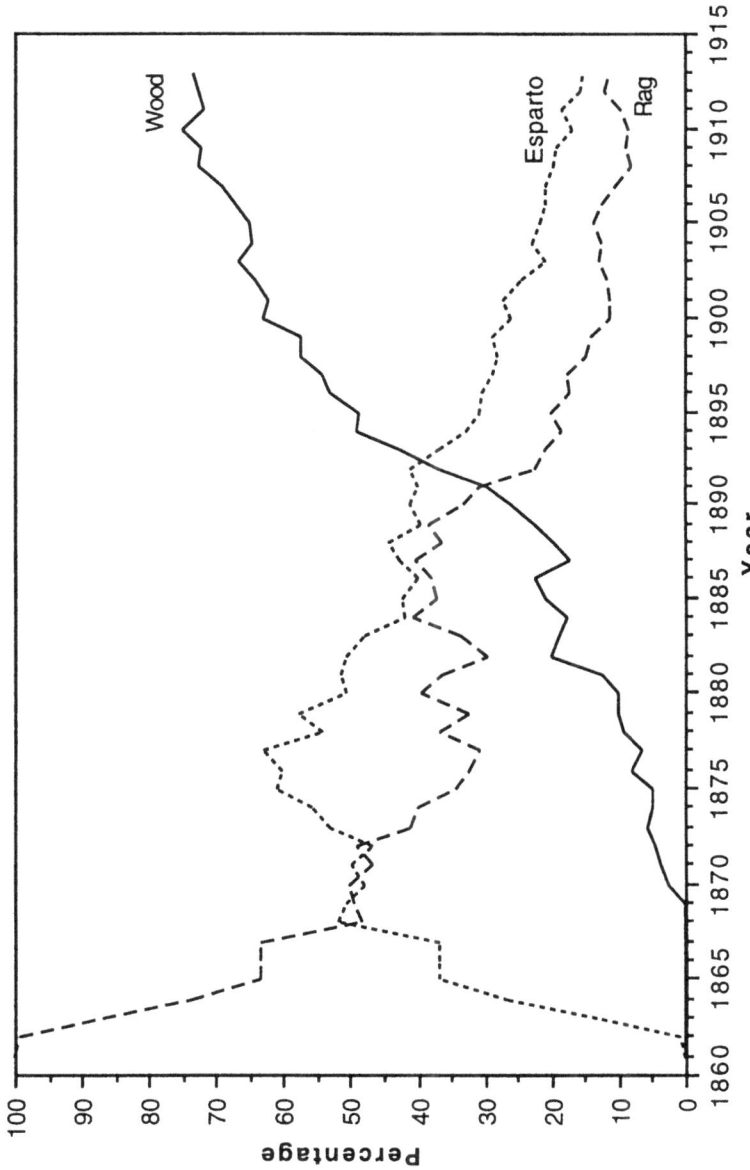

4.1 Raw papermaking material usage in Britain, 1861–1913

Table 4.1 *Total quantity of esparto imported into the UK, 1861–1877*

Year	Tons	Year	Tons
1861	16	1870	104,870
1862	876	1871	144,411
1863	19,326	1872	104,870
1864	43,403	1873	102,649
1865	52,324	1874	119,176
1866	70,041	1875	141,900
1867	55,074	1876	130,891
1868	95,880	1877	175,878
1869	87,442		

Source: *Annual Statement of the Trade and Navigation.*

since in Britain these had not attracted the same degree of attention, research, and crucially, use to render them feasible alternatives to the British papermaker. Wood pulp in this period remained very much the wild and fanciful idea of the impractical mind. Given the fact that British papermakers were still looking for a material that produced paper of equal quality to rag paper or could be blended with rags without a major loss of quality, the 'shoddy' paper that mechanically ground wood pulp produced was unlikely to get much of a reception. In a very real sense the happenstance – for that was what it was – that the ability to reduce esparto chemically occurred some twenty years before similarly profitable processes were developed for wood, and that Routledge made his discovery in Britain, a land that was itself bereft of suitable raw materials and was thus not driven to find a way to use a seemingly abundant supply of timber, provided very fertile ground for the rapid implementation, improvement, and spread of esparto as a raw material.

This situation was not to last. From 1878 a steady growth in the use of wood took place. Between this date and about 1889 this rise in importance of wood pulp was primarily at the expense of esparto. In 1877 esparto use stood at 63 per cent of all material usage, but in a mere twelve years this been reduced to 39.5 per cent. Over the same time period wood's share had grown from 6.5 to 22.2 per cent. Interestingly, the position of rag in this second phase seems to have remained fairly unchanged, varying between 30 and 40 per cent. Indeed, if anything its share grew slightly over the period, reaching in 1884 a height of 40.8 per cent that had not been achieved since the early 1870s. The next year, however, saw it fall back to 36.9 per cent. The buoyancy of rags in the 1880s may be due to the fall in its price of about 30 per cent, though esparto prices also fell by around the same amount in this period (as did wood-pulp prices), and its share continued to decline.[154] Esparto's decline seems more closely related to the emergence of the sulphite and sulphate processes in the early 1880s, which revolutionised the production of wood pulp and made it a viable alternative to esparto. Wood could now produce a paper comparable in quality to that which could be made with esparto. Even mechanical wood pulp, enhanced by similar chemical processes after grinding, became a serious prospect to papermakers. Wood pulp was becoming the substitute for the material that was to supplement the rag; it was not yet the substitute for the rag itself.[155]

The stability of rag usage fell apart after 1889, and its share of total tonnage of all material used in papermaking joined esparto in a long and continuous decline. By the turn of the century it had found its new level at around the 10 per cent mark, although in some years, such as 1908, it fell even below this level to 8.1 per cent. Esparto's decline was equally precipitous, ending up around the 15 per cent mark at the outbreak of the war.[156] By this stage wood usage had assumed enormous proportions. In 1896 it first exceeded 50 per cent and by 1914 was approaching 75 per cent of the total. In tonnage (Table 4.2) the growth of wood pulp was also impressive. From the first year that wood-pulp imports were separately classified in the Board of Trade returns in 1887 to 1913, wood pulp use grew at an average compound rate of 10.13 per cent. Almost all of this wood came initially from Norway, Sweden, and Germany, but in the decade and a half before the war Russia and Canada became significant exporters of wood pulp to Britain.

Steady improvement in the manufacture of paper from wood pulp spurred on its ascension to prominence. Information about these improvements was widely available to the papermaker through the industry's press and other publications. For example, in a story on wood pulp in the *Board of Trade Journal* of 1891, it was argued that, 'the fact

[154] The decline in rag prices was largely due to the increased adoption of alternative papermaking materials both locally and overseas. Yet relative prices were probably not that much disturbed by this development. Between 1878 and 1888 rag prices fell on average from £16 to £11; while between 1880 and 1890 esparto prices fell from £7 to £5 and chemical wood pulp from £8 to £5.5. Spicer, *Paper Trade*, pp. 32–3, 39.

[155] 'It must be borne in mind that if we could get rags as cheaply as we can wood it is not likely that we should use the latter by preference ... As we have already stated, wood does not supersede rags, but is only a substitute. For paper-making purposes rags rank higher in excellence than any other known material, and we believe that they will continue to do so yet for a considerable time. In the course of scientific experiments with wood it is much more probable a substitute for that material will be evolved than that rags will be removed from their pre-eminent position.' *PMC*, 10 December 1886, p. 400.

[156] Like rag, esparto did not totally disappear, as the material found a niche for itself in the high quality printing and writing paper sector of the trade. In 1938 316,000 tons of it were still imported into Britain for such purposes. Hills, *Papermaking*, pp. 141–3.

Table 4.2 *UK wood-pulp imports in tons 1887–1913*

Year	Tons
1887	79,543
1892	190,946
1897	388,304
1902	525,799
1907	672,499
1913	977,757

Source: *Annual Statement of the Trade and Navigation.*

that the quality of paper made from pulp in this country has been steadily improving within the past years is sufficient evidence that the various processes are rapidly reaching perfection. It is also becoming possible to use a larger percentage of pulp in the manufacture of paper.'[157] Likewise an article entitled 'Sulphite: Past, Present, and Future' in the *Paper Trade Review* in 1889 not only described the current state of the art with the sulphite process, but conjectured about the even greater achievements to come. Stress, however, was placed on the giant strides already made. It reported that, 'improvements have been made all around during 1888 in sulphite plant both in England and on the Continent, the lead being taken in England by Mr. Edward Partington. The continent of course maintains the lead over England and America, the fertile brain of Dr. Kellner having evolved a series of new processes and improvements.'[158] The Continental lead was attributed to its greater and longer familiarity with the material, although the benefits of this knowledge were obviously not kept from British papermakers. A growing body of information on the various types of wood pulp was thus becoming available to the papermaker which, along with existing knowledge about other materials, permitted ever more rational decisions to be made.

The changing nature of the problem was also reflected in patent applications. Table 4.3 gives an overview of the pattern and composition of British papermaking patents between 1855 and 1913. The numbered columns of Table 4.3 represent (1) the total number of all British patents applied for to do with raw materials and their processing; (2) this number as a percentage of all British papermaking patents; (3) the total

[157] *Board of Trade Journal* 10, no. 54, January 1891, p. 59. Some other references to wood pulp in the same journal are found in 14, no. 18, January 1893, p. 89; 15, no. 106, May 1895, p. 580, and 19, no. 112, November 1895, p. 580.
[158] *PTR*, 1 February 1889, p. 1. Similar articles can be found in all the other trade publications. See, for example, *PMC*, 10 December 1883, p. 272; 11 May and 10 June 1885, pp. 131–4 and 167–70, and 10 March 1887, p. 100.

Table 4.3 *Scale and composition of UK papermaking patent applications,* *1855–1913*

Years	(1) All raw material patents	(2) Raw material patents as a percentage of all paper-making patents	(3) New raw material patents	(4) Known raw material patents	(5) New raw material patents as a percentage of all raw material patents	(6) Known raw material patents as a percentage of all raw material patents
1855–1864	202	59.2	63	139	31.2	68.8
1865–1874	244	66.3	38	206	15.6	84.4
1875–1884	155	48.7	24	131	15.5	84.5
1885–1894	246	51.4	14	232	5.7	94.3
1895–1904	192	37.9	16	176	8.3	91.7
1905–1913	186	36.7	21	165	11.3	88.7

Source: *Chronological and Descriptive Index of Patents.*

number of British patents specifically concerned with finding *new* raw materials for papermaking; (4) the number of patents dealing with the chemical and mechanical treatment of *known* raw materials; (5) number (3) as a percentage of (1), and (6) number (4) as a percentage of (1).

The first thing to note from Table 4.3 is, as expected, the high percentage of the applications made for patents generally to do with the finding and improvement of raw materials for papermaking. Between 1865 and 1874 this represented 66.3 per cent of all papermaking patents applied for by Britons. Over the coming decades this percentage fell to roughly a half of this figure, yet this decline, as columns (1) and (2) show, was neither steady in nature nor attributable to a major fall in the number of patents applied for. The decline proceeded in a steplike fashion; in what appears three phases roughly comparable to the stages apparent in Figure 4.1. This pattern coincides with the process of a gradual narrowing down of feasible alternatives over time, the focusing of effort towards those alternatives, and the growing awareness of papermakers of the options. At the outset of the quest for new raw materials the options open to the papermaker were numerous and knowledge about each limited. As time passed such knowledge was gradually acquired and useful, or at least potentially useful, materials continued to attract the attention of papermakers and inventors, while the less plausible were weeded out. Eventually almost all innovative effort came to be directed towards the fine tuning of existing methods.

Decomposing patents between those which suggested a new raw material and those which suggested an improvement in the treatment of an existing material gives support to this view. As column (5) of Table 4.3 shows, the share of patents concerning new raw materials as a percentage of all British patents to do with raw materials declined steadily between 1855 and 1894 from 31.2 to 5.7 per cent. Correspondingly, the share of patents that involved an improvement to processes applied to existing raw materials rose. This suggests that as time passed less innovative effort was exerted on searching for new materials (column (3)), because more and more innovators came to believe that amongst the materials that had already been suggested lay a number capable of being developed into feasible alternatives. As a result, increasingly more attention was diverted to finding practical and cost-effective means of exploiting these alternatives. These patents involved a motley bunch of innovations, ranging from chemical processes to reduce wood to cellulose, recover waste products, and enhance bleaching, right through to new designs for boilers, pulp strainers, and even improved rag cutters. The thing that linked this diverse collection of innovations together was the desire of their inventors and sponsors to discover means of reducing the cost and augmenting the quality of pulp available to the papermaker by making improvements to one of the known raw materials. As such, these innovations most certainly were a continuation of the search for a staple raw material for the paper industry.[159]

Naturally, the search for new raw materials also continued. There were always those who felt that a perfect raw material was just waiting to be found in some secluded corner of the world, and papermakers could not ignore the possibility that they may be right. In the 1880s a young Frederick Barlow went out to India and Burma to investigate the possibility of using bamboo shoots for papermaking: a suggestion actually made by the discoverer of esparto paper, Thomas Routledge. His report on his return was not favourable, but the fact that he made the trip suggests that papermakers still felt the need to comb the world for materials.[160] The *Board of Trade Journal* also continued to make its

---

[159] Further evidence of the general trend can be found in the relationship between rag usage and patenting activity. Taking rag's share of the industry's raw material usage as an indicator of its progress in finding viable substitutes, one could reasonably assume that the higher that share, the greater the proportion of the industry's innovative effort directed towards raw material related patents. Indeed, the correlation coefficient between the two variables is +0.827 which suggests that the share of patenting activity concerned with raw materials is strongly correlated with the importance of rag to the industry. This high degree of association is also consistent with the theory that the amount of innovative effort devoted to finding rag substitutes in the second half of the nineteenth century was largely influenced by the industry's prior success in this field.

[160] Evans, *Endless Web*, p. 112.

regular reports on possible new raw materials for the papermaker. In August 1890, for example, it quoted the claims of the *South American Journal* of July of the same year that, 'a new factor is entering the papermaker's market. It has been proved that the banana plant contains a greater quantity of pure fibre than any other of the numerous vegetable products used for the manufacture of paper.' It went on to add that, 'the adaption of the plant to commercial purposes will, it is anticipated, revolutionise the paper material market, and largely affect the industry'.[161] Sadly for the banana growers of South America, the claim was unjustified.

In the last twenty years before the First World War there also appears to have been something of an upturn of interest in new materials. This is evident in Table 4.3. In both absolute numbers and in terms of percentages of British patents broadly concerning raw materials, this type of patent showed a definite, if modest, improvement. Bamboo once again came into vogue. Sindall, a chemist and technical adviser to the government of India and Burma, for example, initiated research in 1905 into the feasibility of using bamboo as a source of cellulose. By 1909 Sindall had finished his tests and, realising the value of his findings to British papermakers, as well as Indian, turned his research into a book on the subject. In its preface he made it quite clear that the motivation for his research as well as that of others at this time was: 'the growing need for new papermaking fibres which becomes more acute every year'.[162] Here Sindall was alluding to a lingering fear amongst papermakers that the rapid growth rate of wood-pulp usage of the last twenty years of the nineteenth century could not be sustained and that another material would soon be needed. Fear was also expressed at imagined 'fast failing supplies of esparto'.[163] That these fears were not realised does not mean that they were not real to many in the trade. They were real enough to the famous paper-machine-makers, James Bertram and Sons. So real in fact that the firm, in a move that took it away from its traditional pursuits, established its own laboratory equipped with experimental plant where it produced 'samples of paper from materials suitable for paper-making sent ... from all quarters of the globe'.[164] It

---

[161] *Board of Trade Journal* 19, no. 49, August 1890. The same report was carried in *PMC*, 10 December 1890, p. 470. Other materials suggested included an Algerian grass called *diss* (*Ampelodedmos tenax*), vol. 10, June 1891, pp. 631–2; sugar cane, vol. 14, March 1893, p. 331, and a north Indian grass called *Bhabur* (*Ischoemum augustifolium*), vol. 19, September 1895, p. 332.

[162] R. W. Sindall, *Bamboo for Papermaking* (London, 1909), p. 7.

[163] *Board of Trade Journal* 19, no. 110, Sept. 1895, p. 333. Fear of a depletion of wood supplies were even appearing in the mid-1880s. *PMC*, 10 December 1886, p. 400.

[164] The quotation can be found in an illustrated catalogue of papermaking machinery that Bertrams published in 1921. Bertrams Collection, GD 284/25/9, p. (viii).

was not alone in renewing the search for new materials. In 1911 Dickinsons' Croxley mill imported cotton roots and stalks from Egypt for an unsuccessful experiment.[165] Interestingly, Bertrams also viewed bamboo with great favour. In 1913 it re-issued under its own auspice Sindall's study, adding its own preface to the edition. In that preface the company, concurring with Sindall's belief in the need for a new material, boasted that 'for the past few years we have made a special study of bamboo as a suitable fibre for the manufacture of pulp and paper, and have devoted considerable time and attention to the designing of machinery for the capable and economical treatment of this material . . . There is no doubt that in the near future bamboo will occupy a prominent position as a papermaking fibre.'[166] The point here is not to emphasise the importance of bamboo in the minds of British paper-makers at the turn of the century, rather to show that considerable effort was still being made to find new raw materials by papermakers once again fearful of a depletion of the existing supplies of cellulose. This resulted in an upturn in the number of patents for new raw materials applied for after 1895. This upturn, however, constituted nothing more than a small and transient deviation in the trend towards a general agreement among papermakers as to the handful of raw materials that they believed would dominate the industry in the twentieth century.

Summary

Viewed from the perspective of the British papermaker of the second half of the nineteenth century and with due allowances being made for the sources, timing, and flow of new information about the competing raw materials available, the actual course chartered by the British paper industry with regards to the use of these materials in this period makes a good deal of sense. One must remember the situation papermakers found themselves in in the 1850s and beyond. What confronted them was not simply a choice between fully developed materials and techniques where the essential problem had already been solved and only the optimal blend of the solution needed to be fixed; rather it was a problem that first required a lengthy and uncertain search for possible solutions before such questions of optimality could even be entertained. In that search there was certainly no lack of vigour or purpose shown in Britain. The efforts of Cowans and others considered in this chapter confirm this.

Esparto, even though it could have been more expensive to produce

---

[165]   Evans, *Endless Web*, p. 176.
[166]   Copy found in Bertrams Collection, GD 284/25/11, pp. 5–6.

than mechanical wood pulp, took hold in the 1860s, simply because it was the first and only material in this period to rival the quality of rag paper and meet the requirements of the British papermaker. The initial foothold it gained in the 1860s enabled it to improve and strengthen its position in the 1870s. At that time wood pulp in Britain was still in its experimental stages. By the late 1870s. and early 1880s, however, breakthroughs in the production of chemical wood pulp had been made overseas, and wood pulp began its steady rise to prominence; at first at the expense of esparto, but then later of both esparto and rag. The speed with which wood pulp was adopted once it met the criteria laid down by British papermakers illustrates there are limits to path dependency and that one's initial choice, although important, need not forever be decisive.[167] Moreover, it shows that there was no lack of enterprise in the British paper trade. There were, of course, exceptions, but at every stage, careful consideration, appropriate boldness, and general competence rather than conservatism and failure were the defining features of entrepreneurship in the British paper industry in the late Victorian period. The fact that Britain also experienced considerable export growth and, as we will see in the next chapter, led technologically between 1860 and 1890 when esparto and rag dominated hardly suggests that the industry's choice of these materials unduly impeded its economic performance.

This chapter has also posited a model of entrepreneurial decision-making which differs from most used in the debate on entrepreneurial failure. This model is based on the idea that most decision-making is of an on-going nature, and that information, and as a result decision-rules, are constantly changing. When assessing decisions, it is therefore necessary to consider the quantity and quality of information available at the time the decision is made. In the absence of adequate information intuitive and more 'entrepreneurial' devices play major roles in decision-making. As the information base expands – often as a result of, and in a direction laid down by, earlier decisions – more rationally based decisions become possible. In time when the various options have been more fully worked out the type of optimisation behaviour captured by neoclassical economics becomes a possibility.

One need not accept this model of entrepreneurial decision-making to agree with another argument of this chapter that the questions currently

[167] Contemporaries were well aware of this speed: 'But without exaggeration it may fairly be said that it [wood pulp] has obtained such a hold upon the Paper Makers' minds, and an enthusiasm has been evoked, which never before reached such a climax in the Paper Trade. Considerable interest was shown at the time esparto was introduced into this country, but nothing witnessed like the *furore* at the present time' *PMC*, 10 December 1886, p. 399.

posed in the debate on the role of entrepreneurship in the relative economic decline of Britain are misleading. Prior work on the entrepreneurial thesis, when it goes beyond the purely managerial role of entrepreneurship, invariably views a failure by British entrepreneurs to choose the eventual winner (usually American) as conclusive proof of the entrepreneurial failure thesis. In this chapter it has been argued that this criterion of infallibility is an unreasonable one, unable as it is to take into consideration the context British entrepreneurs found themselves in. A far more reasonable criterion is to demand competence, vision, and vigour, rather than omniscience, from one's entrepreneurs. These need not always bring success – although they will bring their fair share – but they are surely all that a nation can ask from its entrepreneurs.

5   The Anglo-American labour productivity gap

As with many other industries in the late nineteenth century, the American paper industry in numerous ways set the standard to which producers elsewhere aspired. Much about Britain's performance can thus be learnt by comparing its experiences in the second half of the nineteenth century, particularly its productivity history, with that of the American industry. In chapter 3, it will be recalled, our analysis of British productivity growth found that from at least the 1870s onwards American paper manufacturers seemed to have consistently held in terms of labour productivity an approximate two to one lead over their British cousins. This finding appears to be no statistical aberration, as contemporary observers also seem to have been well aware of the gap's existence as well as its approximate magnitude.[1] Analysis by later commentators and modern researchers too has not only identified this same gap, but has also indicated that it persisted and was probably extended well into the twentieth century. Broadberry and Crafts, for example, put the gap in the first half of this century as fluctuating around the two-and-a-half-to-one margin, whereas Rostas in 1935 estimated its magnitude to be of the order of approximately two-and-a-quarter-to-one.[2] These analyses, however, do not venture back into the nineteenth century, so there is little room for direct comparison with our calculations. Broadberry and Crafts do give an estimate for 1907/9 (265.0) which is somewhat higher than ours.[3] This difference is attributable to the fact that they used Fabricant's estimates for the output of the American industry in 1909, while this study has made use of Frickey's estimates. It is believed that Frickey's figures are more reliable since these are based on the paper and forestry industries' own assessments of

---

[1] Tariff Collection, TC7 28/2, p. 2, E(1); V. S. Clark, *History of Manufactures in the United States* (New York, 1929), vol. 2, p. 488; and S. J. Chapman, *Work and Wages* (London, 1904), p. 237.
[2] S. N. Broadberry and N. F. R. Crafts, 'Explaining Anglo-American productivity differences in the mid-twentieth century', *Oxford Bulletin of Economics and Statistics* 52 (1990), 378; Rostas, *Comparative productivity*, p. 36.
[3] *Ibid.*

output, whereas Fabricant's figures are constructed from extrapolations from later censuses as well as estimations from census data of the time which did not distinguish at all between the paper and pulp industries.[4] In any case the story told by both calculations is the same: that productivity levels in America were at least twice those of Britain towards the end of the first decade of the twentieth century and that this differential was not to disappear over the coming decades.

Other recent research by Broadberry suggests that the same phenomenon of a two-to-one productivity lead was also a feature of the relative labour productivity performance of British and American manufacturing as a whole from the mid-nineteenth century on.[5] This is in distinction to comparative GDP per employee figures which, according to some in this on-going debate, show a period of divergence in the third quarter of the nineteenth century followed by one of marked convergence in the twentieth century.[6] On the face of it then, Britain's performance in the manufacturing sector appears far worse than for the whole economy. This is no statistical curiosity, but strikes at the heart of the question of comparative economic performance of the respective countries since the mid-nineteenth century, for it was there, in the manufacturing sector, that Britain's much discussed relative economic decline is supposed to have been located.

General theories explaining the differences between British and American economic performance in the nineteenth and early twentieth centuries, of course, abound. Best known of these is Habakkuk's seminal study of technology on both sides of the Atlantic, which suggested that the differences between the two countries emanated principally from the relative capital-intensity of American manufacturing induced by its abundance of land and scarcity of labour.[7] Since then, various versions of the factor costs argument have appeared in the literature. Broadberry, for example, attributes America's labour productivity leadership in the late nineteenth century and beyond to its more

---

[4]   S. Fabricant, *The Output of Manufacturing Industries, 1899–1937* (New York, 1940), p. 482; Frickey, *Production*, p. 178.

[5]   Broadberry, 'Manufacturing and the convergence hypothesis'.

[6]   For example, see W. J. Baumol, 'Productivity growth, convergence and welfare: what the long run data show', *American Economic Review* 76 (1986), 1972–85; M. Abramovitz, 'Catching up, forging ahead and falling behind', *Journal of Economic History* 46 (1986), 385–406; S. Dorwick and D. T. Nguyen, 'OECD comparative economic growth, 1950–1985: catch up and convergence', *American Economic Review* 79 (1989), 1010–30; and E. N. Wolff, 'Capital formation and productivity convergence over the long term', *American Economic Review* 81 (1991), 565–79.

[7]   H. J. Habakkuk, *American and British Technology in the Nineteenth Century* (Cambridge, 1962). David, *Technical Choice* offers an explanation of how such patterns of technological development became self-sustaining.

widespread use of mass production rather than craft production methods; an outcome largely brought about by the unique resource and factor endowments as well as demand pattern it faced.[8] In a similar tone, others have stressed that it was America's wealth of raw materials that gave it the lead. An embarrassing richness of the materials integral to modern technology and mass production – wood, coal, iron ore, copper, and petroleum – is seen by many students of American economic history as having given American industry a vital economic advantage over its less favourably endowed competitors.[9] In another strain of argument, the existence of America's vast markets, laid open for the first time by the transport revolution of the nineteenth century, is seen as bestowing the American manufacturer with enormous and unprecedented benefits in terms of scale economies.[10] Others, however, have seen the multi-unit, vertically integrated, hierarchical, and managerial firms, that these enormous, wealthy, and protected markets spawned, as the critical and distinguishing features of the American experience.[11] Finally, there is the school of thought that attributes America's relative success since the nineteenth century to the fact that its workers were less unionised and less tied to restrictive craft-based traditions, and hence less able than their British counterparts to resist technological and managerial change.[12]

The presence of the two-to-one productivity differential in the paper industry, therefore, affords us an unique opportunity to look inside this phenomenon and its causes, to consider some of the theories purporting to explain America and Britain's different experiences, and to decide whether such a labour productivity gap was indicative of the general failure of the British paper industry in the late Victorian and Edwardian era. In this and in chapters 6–8 such an analysis is undertaken. In the next section of this chapter the available data on the two industries are examined to determine what they can reveal about the gap. In chapter 6

---

[8] S. N. Broadberry, 'Technological leadership and productivity leadership in manufacturing since the Industrial Revolution: implications for the convergence debate', *Economic Journal* 104 (1994), 291–302.

[9] R. R. Nelson and G. Wright, 'The rise and fall of American technological leadership: the postwar era in historical perspective', *Journal of Economic Literature* 30 (1992), 1931–1964; N. Rosenberg, *Technology and American Economic Growth* (New York, 1972), pp. 18–24.

[10] P. Temin, 'The relative decline of the British steel industry, 1880–1913', in H. Rosovsky (ed.), *Industrialization in Two Systems* (New York, 1966), pp. 140–55; Rostas, *Comparative Productivity*; M. Frankel, *British and American Manufacturing Productivity* (Urbana, 1957).

[11] A. D. Chandler, *Scale and Scope: the Dynamics of Industrial Capitalism* (Cambridge, MA, 1990).

[12] W. Lazonick, 'Production relations, labor productivity, and the choice of technique: British and U.S. cotton spinning', *Journal of Economic History* 41 (1983), 491–516.

the importance of labour and shopfloor practices is discussed and is followed in chapters 7 and 8 by detailed examinations of the productivity gap between 1860–90 and 1890–1914 respectively.

Machine capacity

The first step towards attaining some type of understanding of the productivity differential is to see whether the available data can provide any insight into its composition. Table 5.1 and 5.2 compare some aspects of the two country's paper industries. In Table 5.1 the comparative annual levels of output per machine and output per inch of paper-machine in the US and UK are calculated. For the years for which we have data the picture given by both measures is similar. From the 1890s, or at least 1900, the average capacity of American paper-machines overtake those of Britain, establishing approximately a 50 per cent lead by the end of the first decade of the new century. A rising output/machine ratio is brought about by processing more pulp per year than before. This, in turn, can be achieved by (1) utilising a different and more efficient type of paper-machine; (2) running machinery for longer each working day and on more days of the year; (3) reducing the running time lost on repairs; (4) speeding up the preparation of raw pulp for the production process, so that machine time is not wasted in waiting for pulp; (5) introducing wider machinery; and (6) speeding up the paper-machine itself.

Although there were actually two paper-machines in competition in the industry from the beginning of the century, the mainstay of the paper industry everywhere soon became the Fourdrinier. Factor (1) thus provide us with no difficulties. The tenacious and lasting competition in the United States of the alternative Cylinder machine invented by John Dickinson in 1812, however, is an interesting aspect of the technological history of the industry which is worthy of some mention in passing, not the least because it represents another example of a temporary path dependency. However, as our knowledge of the introduction of paper-machines to the United States is far from complete, these words must necessarily be of a speculative nature.

Despite being a British invention, the Cylinder machine never really caught on in Britain. This was largely due to the fact that its inventor, Dickinson, rather than take out a patent on it, chose to keep it a secret, constructing and repairing all of his machines in his own workshop.[13] Consequently, details of the machine were for a long time relatively

---

[13] MacLeod, 'Strategies for innovation', 293. For a description of the workings of the Cylinder see Clapperton, *Paper-making Machine*, pp. 243–4.

Table 5.1 *Comparative US/UK levels of output per machine year, and output per machine inch per year, 1880/1–1912/14*

Year	Output per machine	Output per machine inch
1880/1	81.3	
1890/1	90.3	
1900/1	120.6	123.2
1907/9	149.3	138.8
1912/14	148.1	136.1

Note: US machine numbers for 1880/1 and 1890/1 are estimated from the 1884 figure.
Sources: Hoffmann, *British Industry*, T54; *Paper Mill Directory*; Frickey, *Production*, p. 16; *Paper Mill Directory of the World* (1900–1914); Cohen, 'Economic determination', p. 4/3; Clapperton, *Paper-making Machine*, p. 247; *Fifteenth Census of Manufactures* (1914), table 24, p. 618; *Census of Production* (1907), p. 624.

unknown beyond the confines of his own mills and with no paper-machine makers initially specialising in its production and development, it was perceived to have little potential.

This story of failure certainly was not the case in the United States where it, rather than the Fourdrinier, became the first machine to replace hand production. Although Thomas Gilpin claimed to have invented and put such a machine into operation at his paper-mill on the Brandywine in Delaware in August 1817, there is little doubt that he had actually got the idea for his device from his brother, Joshua Gilpin, who had visited Dickinson's mills in 1816. For, as the paper historian Clapperton noted, 'the Cylinder paper-machine which was in very general use in America in the latter half of the nineteenth century does not appear to have changed much, if at all, from the original design of John Dickinson in 1812'.[14]

The real breakthrough for the Cylinder machine in America, however, came with John Ames' improvements on the basic design used by the Gilpins. Acquiring the know-how through a classic piece of industrial espionage, John Ames, a papermaker from Springfield, Massachusetts, was able to produce a less complex variant of the Cylinder machine, which was much easier and cheaper to construct. The first Cylinder machine that was commercially manufactured by machinists was made by Phelps and Spafford of Connecticut in 1830, and thereafter it became, unlike in Britain, the main product for paper-machine makers for the next fifty years. In this time further improvements were made to the machine in the United States including its adaptation for the

[14] *Ibid.*, p. 243. See also H. B. Hancock and N. B. Wilkinson, 'Joshua Gilpin: an American manufacturer in England and Wales. Part II.' *Newcomen Society for the Study of the History of Engineering and Technology Transactions* 39 (1960–1), 66.

manufacture of multi-ply paper and board in 1863; an innovation that later found acceptance among British boardmakers and which allowed the Cylinder to survive in the British industry, albeit in a different role from that in which it had started.[15]

By contrast the Fourdrinier's introduction to the United States was relatively slow. Indeed, it was not until 1827 that Henry Barclay installed one which he had imported from Britain in a mill owned by Beach, Hommerker, and Kearney in Saugerties, New York.[16] Although there is a record of an American machinist producing a Fourdrinier in 1829, for many years most Fourdriniers in America were constructed and serviced with parts imported from Europe. Whilst the American industry's seeming lack of interest in a device that was literally revolutionising the industry in Europe may with the benefit of hindsight be regarded as somewhat perplexing, American manufacturers at the time appear not to have been too concerned with the matter. The explanation that was most commonly given for their unwillingness to embrace the Fourdrinier in the first half of the nineteenth century was that it cost much more to import and assemble than the Cylinder machine. Needless to say this was presumably because machinists were only familiar with the workings of the Cylinder machine; a claim evidenced by the fact that early American paper-machine makers seemed to have had difficulty constructing a cheap and effective shaking mechanism – an integral component of the Fourdrinier paper-machine.[17]

Nevertheless, after the Civil War, and especially from the 1880s onwards, the Fourdrinier machine did start to make its presence felt, becoming the pivotal machine by the turn of the century. Indicative of this trend was America's 'paper city', Holyoke, Massachusetts, whose first Fourdrinier was introduced in 1843. By 1897 95 per cent of all paper-machines in the town were of this kind.[18] The eventual dominance of the Fourdrinier is also reflected in the fact that the historical description of the industry given in the United States' *Twelfth Census of Manufactures* in 1900 emphasises the importance of the Fourdrinier, mentioning the Cylinder only once in passing.[19] The current paucity of material on the history of American paper-machines necessarily makes any attempted explanation for this apparent turn-around tentative by nature. Yet the information available is consistent with the view that the Fourdrinier's late success in America was at least in part tied to its

---

[15] J. A. McGaw, *Most Wonderful Machine: Mechanization and Social Change in Berkshire Paper Making* (Princeton, 1987), pp. 101–2.
[16] M. Keir, *Manufacturing* (New York, 1928), p. 482.
[17] McGaw, *Most Wonderful Machine*, p. 102.   [18] *Ibid.*, pp. 161–2.
[19] *Twelfth Census of Manufactures* (1900), pp. 1023–7.

manufacturers' belated realisation of the cost advantages of the Four-drinier *vis-à-vis* the Cylinder; a fact which was, in turn, probably attributable to its paper-machine makers' lack of familiarity with the Fourdrinier until the mid-century. In particular, the Cylinder machine was disadvantaged by the inherent speed limitations of its construction. As papermakers became ever more aware of the need to speed up production, they soon discovered that as the speed of the Cylinder machines increased, so did the centrifugal force which threw stock off the rotating wire mesh drums.[20] This fact alone suggests that even if the American industry did make more use of the Cylinder machine than the British industry – which also used it – this would have been to the detriment of its output/machine ratio rather than its advantage. Machine differences cannot, therefore, account for America's greater average machine capacity at the turn of the century.

Factor (2) is also not likely to have had a major influence in these calculations, for by these dates both industries were running their mills around the clock and for approximately the same number of days each year. Moreover, the effects of temporary slackenings in demand and the trade cycle also appear to have had little impact on the running time of machines.[21] Of the remaining factors that in theory influence average machine capacity, trade journals and contemporary accounts almost all view the speeding up and the widening of the paper-machine as being by far the most important.[22] As we found in chapter 2, it was precisely through these channels that technological change manifested itself in the nineteenth-century paper industry. Moreover, any improvement made to machine capacity as a result of factors (3) and (4) likewise could only be achieved through technological progress. For these reasons Table 5.2 may reasonably be taken to indicate that the rate of technological progress in American paper machinery exceeded that in Britain: a finding corroborated by the comparative total productivity calculations presented in chapter 3.[23] More specifically, it would seem to confirm the lead of American paper-machines over British noticed by contemporary observers from the late 1890s on. These observers all saw American

---

[20] Cohen, 'Economic determination', p. 4/38.
[21] In the 1900s workers in both countries worked on average about 58 hours per week and on 325 days each year. See for example, Report of an Enquiry by the Board of Trade into the Earnings and Hours of Labour of Workpeople of the United Kingdom, *PP* CVIII (1912/13) (hereafter *Earnings and Hours*), p. 310; *RC Labour*, pp. 507–11; *Fourteenth Census of Manufactures* (1909), p. 752; and Dept. of Labor, *First Annual Report of the Commissioner of Labor* (Washington DC, 1886), pp. 388–91.
[22] In Spicer, *Paper Trade*, pp. 69–70 it is estimated that they account for 85 per cent of the increase in output since 1800.
[23] This is also the belief of the trade itself. See, for example, U.S. Tariff Board, *Pulp and Newsprint Paper Industry* (Washington DC, 1911), p. 53.

machinery as larger, quicker and more efficient than the equivalent equipment back home in Britain. As Dyson of the Amalgamated Society of Paper Makers and delegate on the Mosely Industrial Commission noted in 1902:

when we leave the natural resources and go into the mill, there is no doubt we are lagging behind, the mechanical equipment of the American mills being superior to the great majority of the mills in this country, not only in the machinery actually necessary for paper manufacture, but for labour-saving also. The idea of the American is, from the time the raw material enters the mill to get as much of it made into the finished article in the shortest possible time, everything necessary in the manufacture being regulated by this desire.[24]

The difference in the two measures presented in Table 5.1 can be accounted for by the fact that output per inch of paper-machine effectively removes factor (5) in that it excludes changes in the width of the average machine from the calculation. It thus incorporates those productivity changes that are independent of the increasing scale of machinery. It also lets us gauge how much of the greater machine capacity was due to larger machinery. After 1900 this was around 10 per cent.

Manning levels

An increasing technological lead from the 1890s, however, does not explain how important that lead was in shaping relative labour productivity levels, nor, more importantly, the existence of that gap prior to the 1890s. Unfortunately, the data only permit average machine capacities to be stretched back to 1880, but the figures for 1880 and 1890 indicate that, if anything, Britain held something of a technological lead in this period and possibly before. The matter, therefore, is obviously more complicated than different degrees of technological advancement. If technological change operates primarily on the numerator of the labour productivity identity, it is legitimate to ask what was happening to the denominator, labour input? Looking at Table 5.2 which compares American and British labour/inch and labour/machine ratios, another aspect of the overall picture emerges. Both ratios show large gaps which converge somewhat from the 1890s. These gaps suggest that for each machine operated in the industry, British paper-makers on average employed significantly more labour. As these figures are supported by the available contemporary evidence on labour/

---

[24] W. Dyson (of the Amalgamated Society of Paper Makers), 'The Paper Trade', in *Mosely Industrial Commission to the United States of America, October-December 1902: Reports of the Delegates* (Manchester, 1903), p. 216.

Table 5.2 *Comparative US/UK labour per machine, and labour per machine inch, 1880/1–1912/14*

Year	Labour per machine	Labour per machine inch
1880/1	41.7	
1890/1	46.8	
1900/1	66.4	67.8
1907/9	66.0	71.0
1912/14	65.9	71.7

Note: The number given in table is the US figure as a percentage of the British figure in any particular year.
Sources: Spicer, *Paper Trade*, p. 146; *Paper Mill Directory*; *Twelfth to Fifteenth Census of Manufactures* (1900, 1909, 1914); *Paper Mill Directory of the World* (1900–1914); Cohen, 'Economic determination', p. 4/3; Clapperton, *Paper-making Machine*, p. 247; *Census of Production* (1907).

machine ratios, it seems reasonable to conclude that less labour was in fact required in the United States than in Britain for machine-made paper.[25]

But how important was this fact? An outer-bound estimate of the impact of higher labour/machine ratios on British productivity can be calculated by employing the assumption that British machines could be run with American manning practices without suffering any loss of output. This is the equivalent of asking British mills to adopt the prevailing American labour/machine ratios along with the conditions that were necessary to achieve these ratios.[26] Obviously these conditions were an important part of the story, but we shall defer our discussion of exactly what they were and how Britain could have emulated them until chapter 7. For the present, our chief purpose is to establish some estimate of their overall importance in creating the observed gap in the levels of labour productivity attained in each country. The counter-factual that is being suggested takes the form

$$(Q/L_{uk})^{\circ} = \alpha_{uk} \times (M/L)_{us}$$

where $(Q/L_{uk})^{\circ}$ is the estimated level of labour productivity in Britain once American labour/machine ratios are adopted, α_{uk} is capital productivity in Britain, which in each period is regarded as a fixed technological coefficient, and $(M/L)_{us}$ is the machine/labour ratio in

[25] For example, see the NUPMW's reply to the official questionnaire in *RC Labour*, p. 755.
[26] In other words, the amount of labour used in Britain would be $M_{uk}(L/M)_{us}$ or the number of British machines (M_{uk}) multiplied by the quantity of labour needed in the US to man the average machine $(L/M)_{us}$.

America.[27] These estimated labour productivity figures are then combined with actual British and American labour productivity figures in each period, as in the identity below, to ascertain what percentage of the US/UK labour productivity gap can be explained by British labour/machine ratios.

Percentage explained $= \{(Q/L_{uk})^{\circ} - Q/L_{uk}\}/\{Q/L_{us} - Q/L_{uk}\} \times 100$

In this identity, $(Q/L)_{uk}$ and $(Q/L)_{us}$ are the observed labour productivities in Britain and the United States respectively. Table 5.3 gives the results of this counterfactual.

Although the available data are limited and patchy, the findings of Table 5.3 do suggest strongly that prior to 1890 all of the productivity gap between America and British papermaking could be explained simply in terms of labour employment. Furthermore the fact that the counterfactual more than explains the gap also suggests that capital productivity in Britain in this period was not only not lagging behind, but probably exceeding the levels attained on the other side of the Atlantic. Yet another interesting aspect of the calculations is that they appear to suggest that it is only after 1890 that labour/machine ratios are unable to explain all of the observed labour productivity gap. This, of course, as noted earlier, was the period in which America started to establish a technological lead in the field. Combining these two findings we start to get a picture of the forces sustaining the two-to-one labour productivity differential. This is a picture that concerns itself with the interaction of several concurrently active factors. In outline this explanation attributes lower productivity in Britain chiefly to factors influencing the employment of labour in the mills up to 1890 and to a less rapid improvement in paper-machines after that date. This is not to say that labour/machine ratios ceased to matter in the twentieth century, only that they alone cannot provide a complete explanation of the productivity differential from that time onwards.

Before proceeding further a few assumptions implicit in the above counterfactual need to be addressed. They are: (1) that any differences in the levels of human capital that existed between Britain and America could not have affected the outcome of the above counterfactual; (2) that both countries operated similar machinery; and (3) that differing rates of vertical integration did not bias the results. These are all strong assumptions to make, and we will need to consider the validity of each *seriatim*. As for (1), we could expect human capital to affect the counterfactual by reducing the number of employees needed to run each

[27] The equation is derived from $(Q/L_{uk})^{\circ} = Q_{uk}/\{M_{uk}(L/M)_{us}\}$.

Table 5.3 *Estimated UK labour productivity levels using US labour/machine ratios, and the percentage of the US/UK productivity gap this can account for, 1880–1914*

Year	$(Q/L_{uk})^{o}$ (tons/year)	Q/L_{uk} (tons/year)	Q/L_{us} (tons/year)	Explained (%)
1880/1	20.33	8.47	16.53	147.15
1890/1	29.79	13.95	26.89	122.41
1900/1	30.52	20.27	36.81	61.97
1907/9	30.96	21.97	46.21	37.09
1912/14	33.39	23.94	49.45	37.04

Sources: As for Tables 5.1 and 5.2.

of the processes; the underlying assumption being that as a worker acquires relevant knowledge, he becomes better able to handle his task and run his machine with less help from others. The greater endowment of human capital then, the less the number of employees the mill needs to hire. Although there are frequent references to the better education of American workers after 1890, there is certainly no indication of any noticeable differences at earlier dates. Moreover, these claims that American paper-mill workers were better educated than their British counterparts also invariably note that British workmen were equally competent in carrying out their work.[28] As we will find later in chapter 8, America's advantages in terms of the basic educational level of its labour force aided it through its spur to technological change and not through its substitution for proportionately more less-skilled workers. The room for such substitution in any case in the paper industry was particularly small. The nature of the production process, the enormous size of the paper-machine, and the ever-increasing speeds and widths of the machine made it a physical impossibility for one particularly knowledgable worker to replace several less capable ones; unless, of course, deliberate overmanning was being practised.[29]

If different production technologies, each with its own distinct manning requirements, were used in each country, this might also invalidate the above counterfactual. Luckily by this stage, as was discussed earlier, there was a remarkable degree of homogeneity in the paper machinery used. In any case the labour requirements of the Fourdrinier and the Cylinder machine were almost identical, so that this factor was unlikely to have affected labour/machine ratios.

[28] For example, see Dyson, 'Paper Trade', p. 218.
[29] Cohen, 'Economic determination', p. 3/28.

Differences in the machinery employed by American and British papermakers, therefore, provide no complication for the results obtained above.

A more advanced degree of vertical integration may distort the results in that through the expansion of papermaking companies into the marketing of their own products and the supply of their own raw materials a higher labour/machine ratio may be attained. This is undesirable for our purposes as such a rise may give the false impression that labour/machine ratios in the two countries are converging because practices are becoming closer in both when in reality it is actually due to the fact that we are no longer measuring the same thing in each country. If this were the case, we would expect *ceteris paribus* that when integration became an issue it would increase the number of labourers employed in the average mill, so that the industry with the higher degree of vertical integration should have the higher labour/mill ratio. Unfortunately, we have only very patchy data for the labour/mill ratio in the United States. We are on better ground when it comes to the number of establishments operating in America. Each American census of manufactures recorded the number of establishments in the industry rather than the number of mills. An establishment was defined as all the mills held by the same owner in a particular region.[30] If we calculate the average number of employees per establishment in the United States and compare this with the average number of employees per mill in Britain, the result will be biased against the United States; that is, labour per establishment is bound to be greater than labour per mill. These results are shown in Table 5.4. Both the figures using establishments and those using actual mill numbers demonstrate that even in the twentieth century when the United States was arguably more advanced in terms of integration, the amount of labour employed in British mills was significantly greater than in the average American mill of the time. Moreover, the scale of output produced in the average mill in both countries was fairly similar throughout the entire second half of the century, with the American mill on average being at the most 10 per cent larger than the British. These findings suggest that differing degrees of vertical integration alone cannot explain different labour/machine ratios. Indeed, the fact that the ratio is considerably lower in the United States at all times implies that even if vertical integration is a factor, its influence is always greatly outweighed by some other factor that affects employment practice in the two countries.

[30] *Twelfth Census of Manufactures* (1900), p. 1015.

Table 5.4 *Comparative US/UK labour per mill/establishment and labour per mill, 1860–1912*

Year	Labour per mill/establishment	Labour per mill
1860/1	40.31	31.41
1870/1	38.59	31.99
1880/1	50.80	35.90
1890/1	49.16	
1900/1	57.16	43.40
1907/9	62.58	
1912/14	60.88	

Sources: *Paper Mill Directory*; Spicer, *Paper Trade*, p. 146, appendix 5; *Eighth to Fifteenth Census of Manufactures* (1860–1914); Munsell, *Chronology*, pp. 224, 233, Mulhall, *Dictionary*, p. 786.

Machine speeds

Another source of evidence that supports our explanation is the information available in the literature and trade journals on maximum operating speeds of paper-machines in Britain and the United States in different periods. This is useful information, because, as we have seen, the major manifestation of technological progress in the paper-machine in the latter half of the nineteenth century was in increases in the operating speed of the machine. In Table 5.5 this information is summarised. It suggests that up to the late 1880s British and American maximum speeds were comparable, with, if anything, a slight lead being held by Britain. In the 1890s, and certainly by 1900, American maximum machine speeds shot ahead of British, growing at an average compounding rate of 5.4 per cent per annum compared with 3 per cent in Britain in that period. Relatively faster growth rates were also experienced in first half decade or so in the new century. By this stage, it was widely recognised in the trade that American machines set the example in speeds to be followed. Caleb Waite, an employee of Edward Lloyd Limited, sent to the United States by the firm in 1897 to look into the possible purchase of American paper-machines, expressed amazement at the rate of improvement in speeds that had been made there.[31] Table 5.5, therefore, matches fairly closely with the periodisation outlined above where it was also argued that American innovation in paper machinery only began to exceed British for certain in the 1890s.

Of more relevance than maximum speeds to our study of productivity

---

[31] *PMC*, 10 July 1897, p. 326. See also *PMC*, 10 September 1897, p. 440; Clark, *History of Manufactures*, vol. 2, p. 488, and Dyson, 'Paper Trade', pp. 216–18.

Table 5.5 *Maximum operating speeds of paper-machines in the US and UK,
1862–1911 (feet per minute)*

Year	US	Year	UK
1867	100	1862	100
1872	175	1871	175
1887	250	1885	300
1900	525	1900	480
1905	618	1907	550
1911	700		

Sources: Spicer, *Paper Trade*, pp. 66–7; Hills, *Papermaking*, p. 158; Munsell, *Chronology*, p. 219; Weston, *Chronology*, p. 7; Cohen, 'Economic determination', p. 4/26; Hunter, *Papermaking*, pp. 576–8.

is average operating speeds. These are obviously harder to get information on in the contemporary literature. From 1890 on, however, there are sufficient data available for estimates of average machine speeds in both the United States and Britain to be made. This can be done by using the following identity

$$Q = M \times H \times W \times S$$

where Q is annual output of the industry in square feet; M is the number of paper-machines operating in the industry in that year; H is the average number of hours that machinery is run per year; W is the average width of the machines in feet and S is the average speeds of the machines in feet per hour. Taken together H, W, and S are the same as the machine productivity ratio calculated in Table 5.1. If we have values for Q, M, H, and W, then by rearranging the above identity, we can calculate S, the average speed, as a residual.

Of the data needed, the only piece that is not available is the average number of hours machinery is run per year in the industry. This involves two aspects: (1) how long it is run each day; (2) how many days it will be run per year. As for the first, we know that papermakers had strong incentives to run their machines around the clock and generally did, barring downtime needed to make repairs and change wires and felts. They also hoped to avoid unscheduled shutdowns due to insufficient orders or labour unrest. These were less easy for the papermaker to make allowances for, but given the buoyancy of demand in this period and the impotence of trades unions in the machine-mills in both countries, temporary closures forced on mill owners were unlikely to be excessive or differ significantly between the United States and Britain. In Britain, where the pinch of foreign and domestic competition was perhaps more acute, there are very few accounts of mills and their machines remaining inoperative or on half time for long periods. Testimony given to the Tariff Commission in 1904 confirm this observation.[32] Faced with this evidence, a parsimonious estimate of *H* between 1890 and 1914 would be 6,000 hours of running time per year. This is based on the assumption that mills in both countries ran 18.5 hours per day for 6.5 days of the week; a figure that allows 5.5 hours of downtime for repairs and other factors on average each day. This is the equivalent of 120 hours a week for 50 weeks of the year, or 6.5 weeks of idleness. Stoppages are unlikely to have been greater than that for the industry as a whole.

Another problem is that our annual output figures need to be converted to square feet from tons. This has been done by estimating a conversion rate from information available at the time. For example, Kay in 1893 tells us that the 139 miles, 874 yards of 4 foot wide standard printing paper produced at the Star Paper Mill, Fenniscowles in one day in that year weighed some 10 tons 3 cwt. This is the same as saying that each ton of paper contained 290,262.86 square feet of paper. Other examples given likewise confirm that a conversion rate of around 290,000 square feet of paper per ton is realistic. Of course, a certain degree of variation is to be expected between the different grades of paper, although the vast majority of paper made would have conformed fairly closely to this basic rate.[33] In any case, as the composition of American and British paper output in this period was roughly similar, any error resulting from the use of this particular conversion rate is unlikely to affect the comparison attempted. Putting all the information together, we get the results presented in Table 5.6.

The estimates of comparative average operating speeds presented in Table 5.6 confirm the picture given by the information on maximum speeds and the earlier counterfactual. From a position in 1890, where average speeds are similar in both countries or even greater in Britain, American average speeds begin to surpass those achieved on average in Britain. This was noted by contemporaries for all grades of paper. As Dyson remarked about American paper-machines in 1902 'even in higher grades of paper, such as book, bond and ledgers, the machines are built with the same advantages mentioned in the news machines, and do run quicker than the machines in this country making the same class of paper'.[34] It should be noted that these calculations of average speeds will be affected by error in our estimate of *H*, in that, if it exceeds the real

---

[32] See comments of Evans, Nuttal, Dixon, and Garnett in Tariff Collection, TC3 1/84, p. 12; TC3 1/85, p. 6; TC3 1/86, p. 11; TC3 1/87, p. 2.
[33] Kay, *Paper*, p. 61.   [34] Dyson, 'Paper Trade', p. 216.

Table 5.6 *Average operating speeds of paper-machines in the US and UK, 1890–1914*

Year	S_{us}(ft/min)	S_{uk}(ft/min)	$(S_{us}/S_{uk}) \times 100$
1890/1	97.68	103.57	94.31
1900/1	202.51	162.66	124.64
1907/9	291.10	208.02	139.94
1912/14	333.97	245.14	136.24

Sources: Hoffmann, *British Industry*, T54; Frickey, *Production*, p. 16; *Paper Mill Directory*; *Eleventh to Fifteenth Census of Manufactures* (1890–1914); *Paper Mill Directory of the World* (1900–1914).

average number of hours worked, it will effectively underestimate the true average speed. This sort of error, however, is not likely to be great, as our results are not particularly sensitive to minor changes in *H*. For example, an error of 24 hours either way in either country in 1912 will alter our calculations by one-third of a per cent at most. Even an error of one entire working week will change our average speeds estimates by 2 per cent at most. This holds true for our comparative results as well.

Summary

The available data considered in this chapter suggest very strongly that before 1890 American paper machinery held no technological advantage over Britain and that its higher productivity in these years can be attributed to its lower labour requirements. The situation, however, was not to remain static. By the last ten to fifteen years of the nineteenth century American paper-machines began to bypass those operating in Britain both in terms of width and speed. Indicative of the interest this efflorescence of American technological ingenuity in papermaking generated in Britain is the emergence of regular sections in the British trade press on the progress of the industry in America. As early as 1883 the *Paper Maker's Circular* replaced its infrequent reports on America with its regular feature, 'American Gleanings', because its editors had already come to realise that '... so much importance [was] attached to paper manufacture on the other side of the Atlantic'.[35] In chapters 6, 7, and 8 our attention is directed towards finding explanations for Britain's relatively high labour requirements and America's late technological surge.

[35] *PMC*, 10 December 1883, p. 281. Also see *PMC*, 10 November 1886, p. 374 where an American expert claims that American papermaking is now ahead of European and British.

6 Unions and manning practices in Britain and America

In chapter 5 it was argued that, relative to the United States, British paper-mills employed significantly more labour per paper-machine installed. This was most markedly so between 1860 and 1890. In trying to explain this fact, perhaps the best place to start is to determine whether differences in the attitudes of the workforce in each country could have significantly influenced respective practices with regard to the numbers employed per machine or per mill. A major theme of the literature is the hostility of British workers and their organisations to changes that altered the existing customs and arrangements regarding employment in their industry. This was usually a defensive action against technological change that saved on labour and compelled those left to work longer hours for similar or less pay.[1] But these changes were resisted not only because they reduced labour requirements within the workplace, but because they were also often deskilling by nature.[2] In industries where there was a very strong craft-based tradition, the dilution of the trade with the introduction of unskilled labour to operate such new machinery was a tendency that was fervently resisted. This was especially so, because in late Victorian Britain significant social status was ascribed to craftsmen. For their part craftsman revelled in this status, generally regarding themselves as a cut above other workers. As Harrison and Zeitlin noted:

The dignity and pride which they exhibited in confrontations with their employers was matched by the profane indifference which they often displayed towards the fate of their 'inferiors' particularly women and children. If they sometimes exhibited a heroic solidarity with their fellow workmen, it was almost always done within the discreetly regulated limits of 'the trades': within the confines set by those identified as 'artisans of our class'.[3]

[1] W. Lazonick, 'Industrial relations and technical change: the case of the self-acting mule', *Cambridge Journal of Economics* 3 (1979), 231–62; Coleman and MacLeod, 'Attitudes', pp. 588–611; W. Lewchuck, *American Technology and the British Vehicle Industry* (Cambridge, 1987).
[2] M. Berg (ed.), *Technology and Toil in Nineteenth-century Britain* (London, 1979), pp. 25–6.
[3] R. Harrison and J. Zeitlin, *Divisions of Labour: Skilled Workers and Technological Change in Nineteenth-century England* (Brighton, 1985), p. 15.

One strategy of the craft and trade organisations that hoped to reverse these trends was to force employers to maintain traditional manning practices in the workplace both in the sense of numbers employed and their training. According to this theory, such attempts were more successful in industries where skilled craft labour could not be easily replaced and was relatively better organised than its employers. Where these conditions did not hold, however, such pressure for traditional manning practices only served to force the employer's hand and actually expedite the disappearance of customary practices.

One possible explanation for the greater labour requirements in British paper-mills, therefore, may be that overmanning was being systematically practised there. The aim of this chapter is to examine this proposition. As we shall see, although there was certainly a strong craft tradition in hand papermaking that had not totally disappeared by the second half of the nineteenth century, there is no evidence at all to suggest that machine-mill workers and their organisations forced the overmanning of machinery in their mills, let alone that such practice was what distinguished the British industry from the American. Rather, what evidence there is suggests that if this practice was significant anywhere, it was more likely to be in America than in Britain.

British unions

Although local combinations and guilds of papermakers had been active in the late eighteenth century, the first proper union to represent the trade was formed in 1800.[4] It was called the Original Society of Paper Makers (OSP), was based in the Home Counties, and, naturally enough in those days before the paper-machine, represented the interests of the entire trade. This, however, was not an organisation that was open to all associated with the paper trade; it was restricted to only those who were 'carded', viz., had done a seven year apprenticeship with a recognised papermaker.[5] It was thus in essence a craft organisation. Because of the dominance of the hand-made trade in the early years of the nineteenth century, as an organisation it wielded quite significant influence, and amongst its ranks it could number virtually every single vatman, coucher, layer, and finisher in the British hand trade.

The spread of the paper-machine, however, undermined its power, because the OSP refrained from becoming involved in that part of the

4   Coleman, *British Paper Industry*, p. 262; C. J. Bundock, *The Story of the National Union of Printing, Bookbinding and Paper Workers* (Oxford, 1959), pp. 356–68.
5   BLPES: Webb Trade Union Collection, Coll. E, Sec. B, LXXIII, item (i), OSP Rulebook, Rule 2, p. 3.

industry. The result was that by the latter half of the century, it, like hand production itself, had faded in importance and ceased to be a concern to the vast majority of those in the industry. The OSP's failure to incorporate machine-mill workers into their ranks emanated from the disdain that traditional hand producers felt for 'shoddy' machine production. It was very much a matter of status, as the OSP regarded those who worked machinery and who had never learnt to make 'real' paper, unfit to share the standing of genuine papermakers. As Bundock wrote, 'there is no obvious reason why the OSP should not have become the representative body of the men in the machine-mills, other than . . . pride of craft'.[6] The desire of the OSP to retain its craft purity was complicated when 'carded' hand-mills converted to machine-mills. Technically, these papermakers who had done the requisite seven years apprenticeship in a hand-mill still qualified for membership of the OSP, yet the OSP was visibly uncomfortable having such machine operatives in their midst. In the 1890s three such mills still existed, and the executive of the OSP reluctantly permitted them to remain in the Society, but reiterated its prior ruling that no more were to be admitted. In Beatrice and Sidney Webb's notes on the OSP made at the end of the last century, it is made clear that the Society still felt that the matter was far from closed:

But it is admitted that the whole question of the machine trade is a very difficult one to them, and they seemed to think that the easiest way out of the difficulty will ultimately be found in getting rid of the few machine men in the Society and making it purely a hand made paper-makers' union.[7]

Not only did the OSP want to keep machine men out of their union, they even fought against working on the product of the machine-mill. At the Phoenix Mill in Dartford in 1884 OSP members went on strike over being asked to apply a finish to a paper that had been made in a machine-mill also owned by their boss. By this date, however, the OSP's strength in such confrontations was not great, and as a result of the strike the mill became a non-Society one. As the *Paper Trade Review* told its readers:

We know for a fact that a strike among paper-makers in the year 1884 is unlikely to affect all the workers in a mill. What paper-makers may do is unlikely to affect the rag pickers and the army of general workers, whose interests are disassociated with the Society's movements.[8]

Moreover, its report went on to say that their places could be easily

6   Bundock, *Story*, p. 368.
7   Webb Collection, Coll. E, Sec. A, XXIX, item 3, p. 244.
8   *PTR*, 5 December 1884, pp. 358–9.

filled by non-Society members or by converting the mill to machine-made paper.[9] The crucial point of this episode was that it illustrated how divisive the issue of hand- and machine-production had become and exactly how, as a result, it had weakened the ability of the labour force in the paper industry to influence their working conditions effectively.

As a result of the OSP's unwillingness to expand so as to encompass at least some of those working in machine-mills, an increasing percentage of workers in the industry found themselves without adequate protection at a time when conditions and wages were under pressure. Into this vacuum stepped the United Brotherhood of Paper Makers in September 1854; an organisation especially set up to represent those with skills in the machine-mill. This included bea-termen, machinemen, and finishers only. In 1869 the organisation split over the issue of tramping, and those who bitterly opposed its continued practice formed the Modern Society of Paper Makers.[10] Both of these organisations which operated side-by-side until 1894 maintained the exclusion of unskilled paper workers.[11] In 1894 the Modern Society and the Brotherhood, after having finally resolved their differences over tramping, reunited to form the Amalgamated Society of Paper Makers. At this stage the new union still only represented the skilled labour force of the machine-mill. Conscious, however, of the need to increase its membership, the leadership of the union in 1907 opted for the admission of half members who paid half the entrance fees and contributions and received half the benefits of full members. This, however, was not to be a way in for the semi- and unskilled mill worker, for a further condition of half membership was that the candidate had to 'possess the same qualifications as at present specified in our rules'.[12] Skilled workmen in the machine-mill thus showed the same prejudice towards the unskilled that hand producers had shown to them. It was not until the formation of the National Union of Paper Mill Workers (NUPMW) on 1 February 1890 that the unskilled in the industry had an organisation that would accept them into their ranks. Its rulebook explicitly opened its membership to 'all male and female

[9] *Ibid.*
[10] Tramping was a practice whereby young and out-of-work papermakers 'tramped' their way between different Brotherhood mills, receiving accommodation and hand-outs from the local branch before moving on to the next mill. Only after a pre-specified route had been completed could the worker return home. This practice, which was originally intended as a method of training young papermakers, by the late nineteenth century had become a means of relieving the local labour market.
[11] See Webb Collection, Coll. E, Sec. B, LXXIII, item (3), Rulebook of the Modern Society of Paper Makers, p. 3.
[12] Bundock, *Story*, p. 395.

workers employed within the gates of a Paper Mill' and pledged itself to the objective of creating a national union that embraced the entire trade.[13]

From its inception the NUPMW initiated a number of attempts to unite itself with the Amalgamated Society. In 1891 it reported to its members that the matter of amalgamation with other papermaking unions had, 'during the last year received the earnest and repeated attention of the head office, not withstanding a singular amount of opposition from the officials of the Modern and the United Brotherhood Societies'.[14] Nonetheless the general secretary went ahead with his proposal and circulated a scheme of dual membership to individual members of these societies, perhaps in the hope that it would kindle some latent desire for unity. However, the proposal seems never even to have got off the ground.[15]

The adoption of the shorter hours movement, that the NUPMW had championed since its foundation, by the Amalgamated Society in the 1890s renewed hopes that through the greater co-ordination of effort the two unions would move in the direction of amalgamation. On 26 December 1896, at an annual meeting of the Lancashire and Midlands District council of the NUPMW, the members of the council unanimously adopted a resolution that read:

That this council expresses its pleasure at what is now being done by the Amalgamated Society of Paper Makers regarding the shorter hours question and requests the Central Executive to pledge the Union to work in harmony and conjointly with the Amalgamated Society in all efforts to secure the Saturday afternoon holiday, and no Sunday night work. Further, we are strongly of the opinion that this can best be done by closer unity between the two societies in the direction of amalgamation or federation, and we trust a scheme for same will be adopted by the delegates at the next annual meeting.[16]

This was duly done, and another letter offering the opportunity for a merging of the two unions was sent to the Amalgamated Society. In response, the general secretary of the Amalgamated Society, whilst acknowledging the need for some co-ordination in their efforts, flatly ruled out any talk of amalgamation.[17] While in Birmingham at the Trades Congress that year, William Ross, the general secretary of the NUPMW, used the opportunity to visit neighbouring paper-mills and tried to create some interest in the union, but met with no success as

[13] Webb Collection, Coll. E, Sec. C, 78, Rulebook of the NUPMW(1890), pp. 5–7.
[14] Modern Records Centre: NUPMW Annual Reports, 41T/BOT, 1891 Report, p. 32.
[15] Webb Collection, Coll. E, Sec. B, LXXIII, item 16.
[16] NUPMW Reports, 41T/BOT, 1896 Report, p. 6.   [17] Bundock, *Story*, p. 384.

nearly all the men he met belonged to the Amalgamated Society and were unwilling to join the NUPMW as well.[18]

In the first few years of the new century serious erosion of the limited inroads that the NUPMW had made in shortening the working week in a few mills prompted Ross yet again to write, though this time with a greater sense of urgency to Dyson, his counterpart at the Amalgamated Society, about the matter of a possible unification:

It was the unanimous opinion of our Executive Council that it would be much better if the Amalgamated would move in the direction of amalgamation or federation with the National Union and thus consolidate their energies and forces. There was no room, nor need for two societies in their trade, and the benefits that would accrue from the merging of the two in one would be almost incalculable. I need hardly assure you that the National Union is ready now as in the past to meet the Amalgamated in conference, and discuss the best means of bringing about either amalgamation or federation.[19]

Dyson's response, however, was not very encouraging:

After thoroughly discussing the suggestions contained therein, we resolved to postpone the question, being unanimously of the opinion, firstly, that the time was not ripe for the adoption of the same, and secondly, that, seeing that there is not more than ten per cent of the workers in the trade who are organised we think there is ample scope for the energies of both societies.[20]

Although the Amalgamated Society did eventually come around in 1912 to co-operate with the NUPMW on the matter of shorter hours, its long reluctance to work closely with the NUPMW seems to have been tied to its fear that co-operation on this issue spelt later amalgamation. As there was also a great deal of friendly agreement between the two unions over the extension of the Merchandise Act and similar issues which did not involve combined effort, there is little doubt that the exclusiveness of the skilled labourer was the main obstacle to unity.

The organisation of labour in the paper industry of the second half of the nineteenth century can thus best be described as patchy and severely divided. The divisions were not only between hand and machine production, but within machine production between skilled and un-skilled labour. Together these divisions acted to weaken any influence the labour organisations could have over the majority of mills in the country.

---

[18] NUPMW Reports, 41T/BOT, 1897 Report, p. 4.   [19] Bundock, *Story*, p. 376.
[20] *Ibid.*

American unions

In America there was not the same record of division. The first serious organisation to appear in the industry was the Machine Tenders Union which emerged in Holyoke in 1884 to protect the wages and jobs of the machine tenders in the industry. As its name suggests, it was essentially a craft-based union that, like the United Brotherhood in Britain, represented only skilled machine workers, was conservative in outlook, and acted in all cases to maintain the existing hierarchy of workers in the mill.[21] The absence of a significant hand-made paper industry in America in the second half of the century put the skilled machine worker at the top of that hierarchy and permitted the labour movement in the American paper industry to avoid one of the divisive issues that racked the British industry.[22] The Machine Tenders Union quickly spread from its base in Massachusetts to all the other papermaking regions of the country, and by 1893 it had remoulded itself as the American chapter of the United Brotherhood of Paper Makers. In that year it was also granted a charter by the American Federation of Labor (AFL). Soon after receiving this recognition, its membership began to dwindle, and its executive realised that if the union were to survive, it would have to extend its jurisdiction into all branches of the trade. In 1897 the American Federation of Labor sanctioned this expansion, and all paper-mill employees were admitted to the United Brotherhood of Paper Makers. The rump of the old Machine Tenders Union in New York did not approve of the dilution of the union with unskilled hands and split from the Brotherhood to establish the International Paper Machine Tenders Union (IPMTU). The IPMTU, however, failed to win AFL recognition, because of the United Brotherhood's prior claims to all paper-mill employees, and could never really gather a momentum that could have swept the 'mixed' Brotherhood away. By 1902 the IPMTU too had come to realise the weakness of their position and returned to the United Brotherhood's fold as members of the newly constituted International Brotherhood of Paper Makers (IBPM).[23]

The merging activity did not end there. The following year the IBPM moved to incorporate all pulp-mill workers into their ranks, in a move once again calculated to broaden the base of the union and to enhance

[21] K. E. Voelker, 'The history of the International Brotherhood of Pulp, Sulphite, and Paper Mill Workers from 1906 to 1929: a case study of industrial unionism before the Great Depression', Ph.D. thesis, University of Wisconsin (1969), p. 29.

[22] C. M. Green, *Holyoke Massachusetts: A Case History of the Industrial Revolution in America* (New Haven, 1939), p. 100.

[23] Brotslaw, 'Trade unionism', pp. 80–6; Voelker, 'History', pp. 29–31.

its bargaining power. At the 1903 Convention of the IBPM, pulp and sulphite workers were duly voted into the union and its membership immediately swelled by 6,000. Pulp-mill workers, however, continued to be organised into separate locals from paper-mill workers and as a minority in the organisation were unable to have much representation on the executive of the Union. Therein lay the seeds of discontent. Led by James Fitzgerald, the pulp-mill workers branch of the IBPM demanded a greater say in the running of the union. As part of this campaign Fitzgerald hoped to unseat Carey, the current president of the union. Both of these figures were apparently dictatorial in manner, and a titanic and bitter clash of personalities, which went beyond the issue of pulp-mill representation, ensued. It came to a head at the 1905 Convention where Carey was successful in getting a motion passed stating that only a papermaker could become president of the union. This was the final straw, and Fitzgerald took the pulp-mill and sulphite workers out of the IBPM and convened a convention in Burlington, Virginia in January 1906, at which the International Brotherhood of Pulp, Sulphite, and Paper Mill Workers (IBPSPMW) was proclaimed. The IBPSPMW comprised all those, including paper-mill workers, who opposed Carey's handling of the affair.[24]

What followed was a brief, but bitter, battle on the shopfloor for supremacy in the paper and pulp industries. The IBPSPMW raided the IBPM's membership and offered special arrangements to any firm that agreed to let the union represent all of its workers. The IBPM's response to this tactic was to go on strike precisely where such an understanding with the management had been reached. For example, at the Fort Edward mill of the International Paper Company (IPC) members of the IBPM contended that several machine workers, who were affiliated with the IBPSPMW, rightfully belonged to them. Because the IBPSPMW refused to relinquish these workers and the company wanted to have no direct involvement in the dispute, the IBPM called a strike. In retaliation, the IBPSPMW local decided that it would do everything in its power to keep the mill running as usual, and assured the management that any place left vacant by a striking worker would be filled by one of its own members. Indeed, the IBPSPMW's determination to break the strike went to such lengths that it even allowed some of its members to operate their machines for 84 hours straight. In the end, the IBPM found it impossible to sustain such pressure and was forced to withdraw its jurisdictional demands and instruct its members to return to work. The real victors in this and numerous other incidents, however, were the

---

[24] Brotslaw, 'Trade unionism', pp. 88–93; Voelker, 'History', pp. 31–6.

employers who took the opportunity in many places to lengthen work hours and reduce wages.[25]

This dispute between the two unions, however, was not about issues of craft and skill, as the IBPM already admitted unskilled labour on an equal footing, but was chiefly political and personal in nature. Interestingly, after Fitzgerald, one of the key protagonists in this inter-union rivalry, retired from public life in 1910, a peaceful compromise and a *modus vivendi* between the IBPM and the IBPSPMW was quickly found and a period of surprisingly close co-operation ushered in. This involved the two unions co-ordinating their organising activities, so that, for example, a practice developed of sending one organiser to a particular plant to form two locals – one for each of the unions. No matter which union employed this organiser, each would share equally in the payment of his salary and expenses. From 1912 they also formed collective bargaining policies and negotiated with management together.[26]

This refound unity of purpose was quickly put to the test by the International Paper Company. In March 1910 the company laid off two members of the IBPM. A workers' committee was appointed to meet and discuss the dispute with the management, whose immediate response was to discharge them as well. On 3 June both the IBPM and the IBPSPMW went out on strike. The strike swiftly spread to other IPC mills where a great deal of animosity to the company and its opportunism over the previous few years still remained. Within a short space of time all of the company's mills had been closed and nearly 6,000 workers were on strike. After eleven weeks the company yielded, realising that the alliance between the IBPM and the IBPSPMW was going to hold, and the unions were granted their terms. These included a three-tour system, a six-day week, a guarantee of non-discrimination against strikers, and a 5 per cent wage rise.[27] It was the greatest victory yet achieved by American paper workers. Such victories were unknown in Britain.

Relative strengths

The greater strength of the unions in America was clearly reflected in the numbers involved in the union movement. According to Voelker, the United Brotherhood claimed that its membership in the United States in the late 1890s stood at around 35,000 which, if true, came to about two-thirds of the entire workforce.[28] Firmer figures can be ascertained from

[25] Smith, *History*, p. 596; Brotslaw, 'Trade unionism', pp. 93–5; Voelker, 'History', pp. 36–47.
[26] Voelker, 'History', pp. 50, 107–8.   [27] Brotslaw, 'Trade unionism', pp. 97–9.
[28] Voelker, 'History', p. 30.

the average yearly membership figures of both the IBPM and the IBPSPMW; these had to be reported to the American Federation of Labor for per capita taxation purposes. Even these figures, however, tend to underestimate the true numbers as they only recorded members who at the time of the report had paid their full dues. Moreover, there was a distinct incentive for unions to understate their membership in these reports, so as to lower the contribution it would have to pay. If we ignore these factors though, in 1905 before the split of the IBPSPMW the International Brotherhood listed itself as having 13,300 members or approximately 19.1 per cent of the entire labour force. According to these same sources, by 1913 the combined number of members in the two unions had fallen to merely 8,000, some 8.5 per cent of all the people working in the paper industry.[29] The heavy loss in members of both organisations seems to have occurred between 1906 and 1910 and have been the result of the conflict between the two unions.

It is harder to get an idea of numbers and pervasiveness of the unions in Britain, not the least because they were so divided. An estimate for 1894 can be pieced together from a variety of sources, which indicates that there were 2,741 paid up union members, comprising 8.6 per cent of the total labour force, in that year.[30] As Table 6.1, which gives NUPMW and Amalgamated Society membership at selected dates, shows, this percentage certainly did not increase in the coming decade. Given the size of NUPMW membership in 1890, as much as 15 per cent of the workforce could have been unionised at that date, although it is hard to imagine this percentage being any higher given that before 1890 the mainly unskilled labourers, who made up the NUPMW, had no organisation that would accept them into their ranks. In any case, at all points it seems that a lower percentage of the workforce in Britain was involved in papermaking unions, and that this percentage probably never amounted to more than 15 per cent of the total.

Unions in Britain, with the exception of the NUPMW, were also less confrontational than their American counterparts and saw their *raison d'être* more in the provision of relief to members who had hit hard times than in altering the conditions that might have inflicted such hardship. Dyson told the American *Paper Trade Journal* whilst there in 1902 that he was struck by the size, breadth, and purpose of the membership of the International Brotherhood in America, and that

---

[29] Brotslaw, 'Trade unionism', pp. 87, 108; Voelker, 'History', p. 4.
[30] The OSP had 640 members, the NUPMW 1,451 and the Amalgamated 650. Interestingly, the overall numbers seem consistent with Dyson's assessment referred to earlier. Webb Collection, Coll. E, Sec. B, LXXIII, item 28, Annual Statement of the Receipt and Expenditure of the Original Society of Paper Makers (1894); NUPMW Reports, 41T/BOT 1894 Report; Bundock, *Story*, p. 380.

Table 6.1 *Membership of the NUPMW and Amalgamated Society of Paper Makers, 1891–1905*

Year	NUPMW	Year	Amalgamated Society of Paper Makers
1890	3,106	1894	650
1891	2,359	1896	800
1892	1,657	1903	1,015
1893	1,649	1904	1,054
1894	1,451	1905	1,095
1895	1,396		
1896	1,292		
1897	1,197		
1898	954		
1899	864		

Sources: Bundock, *Story*, pp. 380, 383, 390, 397; NUPMW Annual Reports (1890–1899).

unions in Britain could not organise on the basis they did in the United States, for in Britain 'the organising of men is on a purely business basis. They join because of the financial benefits they get.'[31] On returning to Britain, Dyson did attempt to widen the scope of the Amalgamated Society, proposing in April 1905 the institution of a society organising secretary who would represent members in disputes with employers or between branches. The proposal, however, was rejected 'without discussion' at the annual meeting after a resounding defeat in a vote by the members, because it was felt that the proposal would be costly to implement and, to many, it smacked of a militancy that they thought was unnecessary, if not destructive, to the membership. As one opponent to the idea claimed:

At the present time, generally speaking, there exists an amicable feeling between Masters and Men which it is undesirable to disturb; but if we appoint a Secretary to go about the country organising and settling disputes, we are afraid that the existing good relationship between Master and Men will be disturbed, also, that in this exists the grave danger of strikes. And as regards the settling of disputes, we are afraid that many of our Masters would not tolerate the interference of a third party.[32]

This was indicative of the attitudes of many in the organisation at the time and restricted the ambit of the organisation immeasurably. At a time when the NUPMW was attempting to pressure individual mill owners by industrial means to institute shorter working hours, the

[31] Reported in *PMC*, 10 January 1903, p. 25; see also *PMC*, 11 April 1903, p. 148.
[32] Bundock, *Story*, pp. 392–3.

Amalgamated Society preferred to petition the Paper Makers Association of Great Britain and Ireland to receive a deputation. When these petitions were casually rebuffed by the employers' association, the Society was in a flounder as to what its next action would be. Invariably it was to send another deputation at a later date. The annual meeting of the Society in London in April 1908 did no more than instruct its secretary to write again to the Paper Makers Association asking for it to reconsider its decision. By this stage some members had grown weary of these tactics and the objectives of the Society that tied the members' hands. As the members from Turner Mills in Yorkshire declared 'it is quite time that the Amalgamated Society of Paper Makers took its place in the Trade Union as a Trade Union, and not merely as a Trade Benefit Society'.[33] At this time, this view was still in the minority.

There is no doubt that the Amalgamated Society's unwillingness to perceive itself as more than a relief society handicapped any aspirations it might have had of influencing work practices in British mills. As such, employers exhibited very little concern about their activities, or just simply ignored them.[34] The managing director of London Paper Mill Company in Dartford, told the Tariff Commission in 1904, that 'the Societies do not govern the rates of wages very much. They appear to be really *bona fide* benefit societies, for the assistance of men in times of need. We have never had a strike that I know of which has been brought . about or helped or instigated by them in any way.'[35] In a similar vein, William Smith, a papermaker from Dewry, who toured the papermaking regions of America in 1894, told trade reporters on his return how much he was struck by the greater strength and organisation of American papermaking unions.[36]

That British unions were patently unable to enforce manning quotas in mills is also supported by contemporary evidence that states that there was no difference between the two countries in the numbers employed to operate each paper-machine.[37] Moreover, there is not even evidence of intent, as no mention in any of the three organisations' rulebooks or records is made about minimum labour requirements per machine or mill, or for that matter about any other aspect of work practice such as machine speeds. Only the Original Society stipulated any apprenticeship restrictions, and these were designed to limit rather than increase the

[33] *Ibid.*, p. 397.
[34] For example, the only reference to the NUPMW in the *PMC* is a very brief article about its desire for shorter hours in 10 July 1894, p. 273.
[35] Tariff Collection, TC3 1/89, p. 1.   [36] *PMC*, 11 June 1894.
[37] This was between five and seven per machine depending on the actual size of the machine. See *PMC*, 10 March 1903, p. 103; Webb Collection, Coll. E, Sec. B, LXXIII, item 18, p. 243; Brotslaw, 'Trade unionism', p. 48.

supply of available labour for use in hand-mills.[38] This was not an issue in machine-mills, since in that branch of the trade apprenticeships had long been done away with.[39] Likewise, there is no evidence in either trade journals, government reports, union archives, or personal papers, that custom and its tacit acceptance by employers exerted an informal and extra-union control over work practices in the industry.[40] Changing speeds and conditions may not have been embraced by all in the machine-mill, but given the weakness of the operative's industrial position in the workplace and the employers revealed desire for such changes, there seems to have been little that opponents of change could have done about it. In any case, the vast majority of those who worked in machine-mills had no experience or affinity with the hand trade and its craft traditions. Indeed, since machine production in the second half of the nineteenth century was still very much in its infancy, these workers even had very little of their own tradition to draw upon. This aside, the tendency alone for work hours and machine speeds to increase significantly and consistently, and with little apparent opposition, over the century clearly implies that if such a tradition actually did exist, it was hardly being respected.

Summary

The evidence that exists suggests very strongly that British papermakers and their unions did not, could not, and probably did not even want to force higher labour/machine ratios or more restrictive work practices in general on employers in Britain than were currently practised in the United States. Furthermore, as there were greater percentages of the workforce unionised, less division amongst the branches of the trade, and a more confrontationalist approach in the United States, a stronger – though equally improbable – case could be made for the contention that American unions actually exerted greater control over their industries' labour requirements than British unions. In this instance at least, then, Britain's lower labour productivity level cannot be ascribed to its pattern of industrial relations.

[38] Webb Collection, Coll. E, Sec. C, LXXIII, OSP Rulebook, p. 78 and Sec. A, XXIX, item 3, p. 204; Fourth Report of the Commissioners on the Employment of Children and Young Persons in Trades and Manufactures not already Regulated by Law, PP XX (1865) (hereafter Fourth Report on Employment), 292.
[39] RC Labour, p. 755; Spicer, Paper Trade, p. 170; Bundock, Story, p. 373.
[40] E. J. Hobsbawm, Labouring Men: Studies in the History of Labour (London, 1964), pp. 344–7.

7   Raw materials, women, and labour-saving machinery: the Anglo-American gap, 1860–1890

In chapter 5 it was found that between 1860 and 1890 the observed two-to-one labour productivity gap between the American and British paper industries was unlikely to have had its explanation in the technological supremacy of the former over the latter. On the contrary, the data available suggest that, if anything, in this regard the British industry at this time was performing better. The source of the problem, rather, appeared to have been the general overmanning of British mills. To many the obvious culprit for such a practice was the greater agitation and resistance to change of British workers of the mid- and late Victorian era. In chapter 6, however, this possibility for the paper industry at least was discounted. To what then can we attribute this apparent over-manning? The aim of this chapter is to tackle this question. The answer it gives sees the overmanning of British mills as the result of two factors: each country's choice of raw material and the greater prevalence of labour-saving machinery in the United States. In both instances it is argued that the America's choice was conditioned by its distinct history and resource and factor endowments. Together these choices caused the labour requirements of British and American mills, especially with regard to women, to diverge from the mid-century; a divergence that continued until 1890. The first section in this chapter concerns itself with the choice of raw materials in the United States in the second half of the nineteenth century; the second considers how these choices affected labour requirements in the mill; while the last deals with the pervasiveness of labour-saving devices in the American industry.

Raw materials again

One of the most pressing problems faced by all nations in the nineteenth century was the failure of the supply of the traditional raw material for papermaking, rag, to keep up with the demand for paper. It was a problem most acutely felt from 1860, and as discussed in chapter 4, the search for its solution became one of the enduring themes of the period.

Although all throughout the world wood pulp was to emerge as the saviour of the paper industry by the outbreak of the First World War, in the interim the response of producers in Britain and America differed. In Britain papermakers until the late 1880s opted almost solely for esparto grass and whatever rag could be found. By contrast, American producers utilised a wider variety of substitutes for rags and allowed rag usage to dwindle far more rapidly than in Britain. While a wide assortment of materials were tried, from the middle of the nineteenth century American papermakers chose to focus their attention on three materials that were initially rejected by the British: straw, cornhusk, and wood.

Of these, straw was the most popular and successful alternative to rag around the mid-century. By the 1820s it had already been commercially used in America in wrapping papers, though its first conversion into printing paper for sale to the public was not till 1829. Printing and writing paper had been made from straw before, of course, on a number of occasions by scientists such as Matthias Koops to demonstrate the possibility of making paper from the material, but till 1829 it had never been commercially produced on such a scale. Adopting a process that the potash manufacturer William Magaw had discovered by accident, G. A. Shryock, a papermaker from Chambersburg, Pennsylvania, in that year abandoned the manufacture of paper from rags and entirely devoted himself to the making of paper from straw.[1] On an average day he produced around 300 reams of mostly printing paper, which he sold for less than $2 per ream; a feat that made the mill, until its destruction at the hands of the Confederates during the Civil War, one of the largest and cheapest in the country. So cheap was the process he used that he was able to produce a copy of the New Testament made from straw paper that sold for a retail price of 5 cents.[2]

This signalled the beginning in the United States of a gradual spread of straw, if not as a substitute, then as a supplement to rag. Improved chemical methods in the following decades aided this spread by enabling straw to be used even in the production of the whitest of papers. By 1849 manufacturers such as Henry Nixon of Springfield, Massachusetts, were making large runs of white paper, usually with the Miller process, almost purely from straw. Mills in other eastern states, especially New York, also opted strongly for straw, so that by 1854 several large newspapers from a number of different states, including Philadelphia's *Public Ledger*

---

[1]   G. A. Shryock, *History of the Origin and Manufacture of Straw and Wood Paper* (Philadelphia, 1866), p. 1.
[2]   Stevenson, *Background*, p. 17; Hunter, *Papermaking*, pp. 394–5, 545; Calder, *First Hundred Years*, p. 16.

and the *Saratoga Whig*, were using paper almost entirely made from this material.[3] The use of straw continued to increase in the following decades, so that by 1884, when a dearth of straw was experienced in New York State, papermakers in Albany agreed to reduce their production temporarily rather than bid straw prices any higher.[4]

Less widespread, but still frequently used before the advent of wood pulp was cornhusk. In 1802 and 1829 American patents were taken out for its use, though the main stimulus to its actual employment in the American industry surprisingly enough seems to have come from Austria. This was principally due to the efforts of the Austrian consul-general in New York in the 1860s, a passionate believer in the material's value to papermaking. With various American sympathisers, he penned numerous letters to key members of the trade and published advertisements and circulars, in which the success the Imperial Printing Office in Vienna had had in using Welsbach's process to manufacture printing paper from cornhusk was described in technical detail.[5] Up to that point its use had been almost solely in the manufacture of various sorts of wrapping paper. By the end of 1863, however, trials with the material carried out primarily in Boston had progressed sufficiently for a printing paper to appear, which even received high praise from the British *Paper Trade Review*. In the following years, the improvements continued and cornhusk soon became reliable enough as a papermaking material for the Associated Press to advocate its use.[6]

Wood pulp, like all the other substitutes for rags, also had a long history of use in the United States. Early and possibly apocryphal accounts put its first use either in 1794 when Matthew Lyon of Fairhaven, Vermont, made paper from rags and the bark of the basswood tree for his newspaper the *Farmer's Library*, or in 1830 when the *Crawford Messenger* in Pennsylvania was allegedly published on wood paper. Both claims, however, have proved hard to substantiate, so that even if it could be shown that wood had in fact been used in these instances, the balance of evidence available

---

[3]   *PMC*, 10 April 1902, p. 139; J. A. Guthrie, *The Economics of Pulp and Paper* (Pullman, 1950), p. 2; L. H. Weeks, *A History of Paper Manufacturing in the United States, 1690–1916* (New York, 1969 [1916]), p. 275; D. C. Vandermeulen, 'Technological change in the paper industry: the introduction of the sulphite process', Ph.D. thesis, Harvard University (1947), p. 33; J. E. A. Smith, *A History of Paper: its Genesis and its Revelation* (Holyoke, 1882), p. 101.

[4]   *PTR*, 11 July 1884, p. 20.

[5]   See, for example, HML: Pamphlet Collection, J. T. Harris, 'Pamphlet explaining the use, benefit, etc. of the products of corn plant; the commercial value of the husks and stalks of corn in the production of fibre for spinning purposes; paper pulp for all description of paper, etc...' (Baltimore, 1867).

[6]   *SC Paper*, p. 291; *PTR*, 1 September 1864, p. 232; Hunter, *Papermaking*, p. 397; Smith, *History*, p. 129.

would still compel one to conclude that they were probably no more than experiments of a very temporary nature .[7] From the 1850s wood pulp began to be utilised on a commercial scale, although it was not until the following decade that its importance started to take on larger proportions.[8] We know that by 1863 the *Boston Courier* had already begun using local wood pulp made by the Voelter-Keller process in its paper. The *Paper Trade Review*'s American correspondent described the process to his readers with amazement:

In colour, in body, and in appearance generally, it bears well upon the rag-made article. Only in shortness of fibre is it unequal to it. The process of reducing the wood into its component fibres is done in regular Yankee fashion, you will say. There is nothing so smart in the old country – not by many chalks. The timber is placed lengthwise in a cylinder with great strength, and it is projected from this with stupendous force by steam power. The blasting it receives from the steam is sufficiently strong to decorticate or disintegrate it into the thinnest threads. So completely is this performed, that the dismemberment of these logs of wood into ribbons of filaments, is more like magic than the work of mechanical power. These fibrous filaments are thereafter easily made into paper.[9]

The results were obviously not too bad either, as producers in Pennsylvania and other Boston papers moved quickly to take up its use.[10] The soda process developed by Hugh Burgess likewise had been introduced early; the first mill being established on 10 acres of land in Manayunk near Philadelphia in 1855. By 1863 this firm had been renamed the American Wood Paper Company, and when its second integrated paper-mill became fully operational two years later, it made a newsprint of a consistency of three parts wood and one part straw from the 12–15 tons of chemical wood pulp it produced daily in its own pulp-mill.[11] Amongst its earliest customers was the *Boston Journal* which by 1863 had shifted entirely to the use of wood-pulp newsprint. Over the coming decade other newspapers made the complete transition to wood: the *New York Staats-Zeitung* in 1868, *New York World* in 1870, *Providence Journal* and *Brooklyn Eagle* in 1871, and the *Albany Argus* and *New York Times* in 1873.[12] In December 1866 the first ground-pulp-mill

[7] *PMC*, 10 August 1895, p. 406; Smith, *History*, p. 128; Hunter, *Papermaking*, p. 380.
[8] Guthrie, *Economics*, p. 3; Clark, *History of Manufacture*, vol. II, p. 36.
[9] *PTR*, 1 May 1863, p. 101. Weston Orr of Troy, New York also used 25 per cent wood pulp in the paper stock for his printing paper in 1853. See Calder, *First Hundred Years*, p. 19.
[10] *PTR*, 2 March 1863, p.75.
[11] A. Proteaux, *Practical Guide for the Manufacture of Paper and Boards* (Philadelphia, 1866), pp. 263–8; Vandermeulen, 'Technological change', p. 53; Clark, *History of Manufacture*, vol. II, p. 132.
[12] H. E. Weston, *A Chronology of Papermaking in the United States* (Wilmington, DE, 1945), p. 6; Hunter, *Papermaking*, pp. 381–2.

in the United States was erected in Berkshire County, Massachusetts, under the guidance of Frederick Wuertzbach, a trained mechanic and woodworker from Magdesprung, and five experienced German wood-grinder operators. Within three months it was making its first shipment of pulp.[13] At first it proved difficult to overcome the prejudice of the producer and consumer against wood. Many newspaper proprietors remained adamantly opposed to wood pulp and either barred it entirely or stipulated the maximum percentage of it which could be used in their paper stock.[14] As Wuertzbach later complained:

I got no encouragement from them. One of the most prominent manufacturers of the time even told me that he would not take any interest in shoddy. Prang [Voelter's agent in America] was told by a news mill manager that as the profits in making news were ample there was no need of introducing the so-called inferior stock.[15]

Once established and demonstrated, however, ground wood caught on rapidly and new pulp-mills and machine shops producing the grinders sprouted up everywhere. By 1870 there were already 11 ground-wood-mills in operation, and 120 of Voelter's patented grinding machines had been purchased in the States since its introduction to the American market less than a decade earlier – some 20 more than had been sold in Germany since 1850. In the meantime local improvement in machinery had made ground wood a much more attractive option to the papermaker. For example, in the 1870s American machinists pioneered the 'hot grinding' process of removing the cellulose for wood by reducing the amount of water used and increasing the pressure applied; an innovation that apparently saved on energy and raised the machine's productivity by over 300 per cent.[16] By 1876 the *Paper Trade Journal* could confidently declare that, as regards to types of raw materials used in papermaking 'in this country wood and straw have the field to themselves'.[17]

In 1860, despite the already common use of straw and cornhusk, rags still dominated raw material usage in the United States, representing some 88 per cent of all raw material used. Still, this was considerably below the corresponding British figure which was practically 100 per cent.[18] Already by this date, however, rags were being used hardly at all in America for the manufacture of the cruder and cheaper varieties of

---

[13] Smith, *History*, p. 132; McGaw, *Most Wonderful Machine*, p. 201.
[14] Clark, *History of Manufacture*, vol. II, p. 485.   [15] Smith, *History*, p. 133.
[16] Vandermeulen, 'Technological change', pp. 45–51; Karges, 'David Clark Everest', p. 21.
[17] *Paper Trade Journal* (hereafter *PTJ*), 11 March 1876. See also *PTR*, 24 October 1884, p. 261; *PMC*, 11 June 1894, p. 277; 10 September 1896, p. 440.
[18] Stevenson, *Background*, p. 18.

paper. Cellulose for paper boards, for example, generally came from a mixture of old papers, fortified with rope cutting or bagging, though occasionally manilla, jute, and straw were also used. These materials, either alone or mixed with waste paper, also provided the stuff used in most varieties of wrapping paper available at that time. Even the vast majority of the newsprint and cheap printing paper was being made from materials other than rag, straw being the single most important. Rag, however, had not totally disappeared from the scene. It remained very much the preferred material for writing paper as well as for the finer grades of printing paper, although some of the medium grades of printing paper and the very finest grades of newsprint had begun to use rag in combination with various proportions of straw, wood, or waste paper.[19] By the 1880 Census when raw material usage was first reported, it is clear that the diminution of the use of rag had proceeded further. Of the total of 619,682 tons of all non-wood raw materials used in papermaking in that year 41 per cent was straw, 14 per cent old paper, 14 per cent manilla stock, 2 per cent other material, and only 29 per cent rags. Unfortunately, these figures exclude wood pulp from their calculation, so the position of rag usage was significantly overstated. Some estimate of wood pulp can be made on the basis of the information that in 1881 1,297 tons of chemical and 3,844 of mechanical wood pulp were being made per day in the United States.[20] If we assume then that 5,000 tons of pulp were made on 250 days of the year – an assumption that probably underestimates the amount of wood pulp actually used – we arrive at an annual figure of wood pulp usage of 1,250,000 short tons. Incorporating this figure with those supplied in the Census, we have the more realistic estimate that wood's share of total raw material usage was approximately 66.6 per cent, straw was 13.6, rag 9.6, old paper 4.6, manilla stock 4.6, and other materials 1 per cent. By contrast the figures for Britain for 1880, which were calculated in chapter 4, put wood at 10 per cent, esparto at 50.4 per cent, and rags at 39.6 per cent. The point of these calculations is that, even if no wood at all had been used in America, rag's share of the total tonnage of raw materials in the States would be below that of Britain.

America's precocity with rag substitutes relative to Britain from the mid-century had its origin in the same pressures and incentives that presented themselves to all papermakers of the period. In America, however, the shortage of rags was even more acutely felt than in Britain. America's population was more scattered; a fact which made efficient rag collection there somewhat more difficult. Lower income levels

[19] Vandermeulen, 'Technological change', p. 35.   [20] Weeks, *History*, p. 290.

combined with a higher cost of textiles also meant that clothes, more frequently than in Britain, tended to be worn in America to the point where they were beyond use to the papermaker.[21] As a consequence America's demand for rag soon far outstripped its domestic supply, so that in order to keep its paper industry from grinding to a halt, it had developed quite a reliance on imported rag. In 1860 almost 65 per cent of the entire paper stock came from domestically collected rags, 12 per cent from cotton waste, rope and bagging, while the rest of the rag material had to be imported to the tune of no less than $1,540,000.[22] Much to the chagrin of the British producer, Britain was the chief source of imported rag. By 1857 rag prices had begun to rise in such an alarming way that papermakers were forced to turn their interest to other possible sources of cellulose.[23]

The Civil War, however, increased this pressure to an unbearable level. As war broke out, the paper industry, which was almost entirely located in the Union, increasingly found it difficult to maintain its supplies of imported rags and cotton waste, because of the blockade imposed on the South, as well as the activities of government contractors who bought up most of the rags for the manufacture of blankets and shoddy.[24] At the same time newspaper circulation grew enormously as readers became eager to hear news of the war and this translated itself into greater demand for paper. Together these two antipathetic forces created a period in the history of the American industry characterised by skyrocketing rag prices and a growing importation of paper, and which the *Paper Trade Review* called the 'American Paper Panic'.[25] The Philadelphia rag merchant, F. A. Server, for example, saw the price of his white rags rise by 348 per cent from 5.75 cents per pound in 1861 to 20 cents in late 1864, and his 'mixed' rags by 300 per cent from 4 to 12 cents over the same period.[26]

In such a climate papermakers were forced to turn to new and undeveloped materials, even if it did mean making a paper that was initially less white, less enduring, and harsher to the touch than pure rag

[21] V. Carlson, 'Associations and combinations in the American paper industry', Ph.D. thesis, Harvard University (1931), p. 302; Vandermeulen, 'Technological change', p. 30.

[22] Weeks, *History*, p. 272.

[23] *SC Paper*, p. 291; McGaw, *Most Wonderful Machine*, p. 191.

[24] F. J. Crawford, 'Manufacturing in the United States, 1860–1870', Ph.D. thesis, University of Wisconsin (1922), pp. 72–103; Proteaux, *Practical Guide*, p. vi; and Carlson, 'Associations', pp. 42, 331.

[25] *PTR*, 2 March 1863, p. 75; 1 May 1863, p. 101; 1 January 1863, p. 39.

[26] HML: Pamphlet Collection, 'Table showing the wholesale prices of articles connected with the manufacture of paper before and during the war in New York and Philadelphia with bills and certificates showing that these prices were actually paid', pp. 4–7.

paper. Unlike Britain, America's fortunate natural endowment offered abundant supplies of alternative materials waiting to be exploited. An abundance of straw awaited the papermaker in the Midwest, providing further impetus for the gradual shift of the industry from New England to Ohio, Illinois, and New York that had begun even before the war.[27] But straw alone was not sufficient to meet the growing demands for raw materials, so that other materials, most importantly wood and corn-husks, were turned to.[28] In time, as more experience and understanding were attained, these new materials became less experimental and the early deficiencies that had plagued them with respect to quality and had provided reason for papermakers not to forget rags, were removed one by one. The fact that these materials also saved on labour, as discussed in the next section, not only expedited their introduction, but ensured their retention after the war was over. Labour costs were after all a particularly important consideration to all American producers in the nineteenth century; a consequence of the abundance of land available for exploitation on the frontier. Virtually all contemporary accounts of conditions in America at the time note that wages there were in general considerably higher than those found in Britain. This was no less true in the paper industry. Dyson, for example, thought that wages were approximately 50 per cent higher in the United States, as did the Select Committee on Pulp and Paper. Likewise a group of American producers, who visited Britain in 1887, also took note that the wages they paid their operatives were at least one and a half times those of the British.[29] Thus in 1893/4, while a machine tender in Britain could make between 44s and 49s 5d per week and a beaterman between 39s and 44s 6d, the same occupations working six days per week in America earned about 72s and 60s respectively. [30]

At the heart of America's greater dependence on non-rag materials in the second half of the nineteenth century then lay the exigencies of the Civil War, which drove rag prices dramatically upwards and which, for the United States, was an accident of history that put it on a new technological path. It was a path that, given the localised nature of innovation, afforded the Yankee papermaker a valuable early familiarity with rag substitutes. Once on that path its resource and factor endowment ensured that there would be no turning back. In the post-bellum period, American papermakers continued to choose the

[27] Guthrie, *Economics*, p. 3.
[28] *PTR*, 1 Sept. 1864, p. 232; Clark, *History of Manufacture*, vol. II, p. 132.
[29] Dyson, 'Paper Trade', p. 215; *Pulp and Paper*, vol. V, p. 3031; *PMC* 10 December 1887, p.432; Carlson, 'Association', pp. 36, 329; *SC Paper*, p. 291.
[30] Spicer, *Paper Trade*, appendix X; *PMC* 11 June 1894, p. 277.

material, in this case wood, that economised on relatively expensive labour.

An interesting question is why American producers largely ignored esparto as a raw material, despite its well-known advantages over straw, at least with respect to the quality of the paper it made: a fact which had weighed heavily in its selection by British producers. It was certainly not due to an ignorance of the material, as the major trade journals of the time carried reports of its progress in Britain. Moreover, as McGaw informs us, the Smith Paper Company of Berkshire, Massachusetts, among others, did experiment with esparto in 1873, but quickly rejected it as being too expensive. This was in spite of the fact that it was reported to have produced a 'beautifully white and fibrous pulp'.[31] The explanation for its rejection by American papermakers lies in factor and resource costs. After paying for the additional labour the processing of esparto required and its transportation across the Atlantic, not to mention the costs of the anticipated shakedown period in which engineers and machinists would have had to familiarise themselves with it, it made economic sense for American manufacturers to stick with rag, wood, and straw. Smith, for example, could conclude from his experimentation with esparto that 'the costs of beating and bleaching it by the only process known' made it as expensive, if not more so, as 'other materials just as good'.[32] In any case, America's abundant supply of straw in the Midwest and wood in the Northeast near to where the industry was predominantly located in this period offered its producers raw materials that were not only cheaper, but as we will see in the next section, saved on relatively expensive labour.[33]

Women and manning requirements

But how important was the issue of raw materials? It will be suggested below that it was very important indeed. Since each material required different quantities of labour to manipulate it into a form suitable for papermaking, it can be seen that the choice of raw material also had a direct impact on labour requirements. This was particularly so when that choice involved the replacement of rag, the material that required the most labour, with some other material. Thus, when non-rag substitutes became available to American producers after the mid-century, a unique opportunity was presented to American papermakers to save on labour and boost labour productivity. As we will see, it was a

[31] McGaw, *Most Wonderful Machine*, p. 198.   [32] *Ibid.*, p. 199.
[33] This was also the view expressed by the American trade journal, the *Paper Trade*, in its *Helps to Profitable Paper Making* (Chicago, 1898), pp. 94–5.

saving in labour that was to fall disproportionately on those who worked in the preparatory stage of production.

With rag paper, the preparation of the paper stock was a complicated and labour-intensive process. The rag arrived at the mill in sacks and bales which were opened and thrown into a duster: a machine that removed the dust and dirt from the garments which in some cases may have been quite old and soiled. After being taken from the duster, the rags were dumped on a table covered with wire gauze. On top of the table stood a long knife that sloped towards the worker, which was used to cut the rag into pieces of about 4 inches. At the same time the worker sorted the rags into different qualities according to type of material, colour, age, and cleanliness. The sorting of the rags was an absolutely essential part of the process, for, unless the rags which were to be beaten together in the engine were all of the same quality both as to substance and condition, the finest and best parts would be ground away in the mill and carried off by the water before the coarser parts were adequately reduced to make a fine pulp. Moreover, the sorters had to remove any buttons and strips of elastic, which may have been mixed with the rags and which, if they had got into the paper stock undetected, would have marred the quality of the paper produced. Once sorted the rags were dropped into small, upright, rectangular wooden boxes that were situated all around the cutting table. From there, these boxes of rags, all of the same grade, were taken to the beater which reduced them to the watery pulp needed to make paper.

All in all, it was a process that required a large number of workers. From Table 7.1, which provides employment figures for an average rag-mill in the United States in 1885, we find that almost 60 per cent of all employees in the mill were involved in the preparatory stage of production, and that exactly 50 per cent were employed in the rag room alone. The most numerous of these employees in the rag rooms were the women who undertook the tiring and tedious tasks of sorting and cutting. A great many of these low skilled workers were needed to keep a mill's machines and vats in constant use. Unskilled male labourers were also needed in the rag-room in large numbers to carry the rags from the storage room to the dusters, from the dusters to the cutting and sorting tables, and then to carry the boxes to the beater. All these workers were supervised by the beatermen who operated and kept the engines running and who were amongst the most skilled and best paid workers in the mill.[34]

As for esparto, straw, and wood, the labour requirements of the

[34] Spicer, *Paper Trade*, pp. 10–11.

Table 7.1 *Number of employees in each occupation in a fine printing paper-mill, Massachusetts 1885*

Occupation	Male employees	Female employees
Finishers	29	12
Machine tenders	11	0
Beatermen	12	0
Rag-room hands	17	52
Repair hands	5	0

Source: First Annual Report of the Commissioner of Labor, p. 389, establishment no. 460.

preparatory process were considerably less than that for rags. Both esparto and straw arrived at the mills dried and bailed up, where they were broken open and sorted through by women who picked out roots and other parts of plant not suitable for papermaking. In comparison to rag sorting, however, this sorting was a much less exacting and time-consuming activity, and for a similar size mill required fewer labourers to attend to it. After the sorting was finished, the material was placed into a willow: a machine which performed a function equivalent to that of the duster in the rag room. From there, the esparto and straw was taken to the engine room, where, as in the production of rag paper, it was transformed into paper stock. The absence of the need to cut the material by hand also reduced the numbers needed to process the raw material.[35]

With wood, the amount of labour required in the preparatory stages of production was even less. For most papermakers of the nineteenth century who used wood, its processing was carried out independently of the paper-mill, and thus added nothing beyond the few unskilled labourers needed to bring the wood pulp into the engine room to the mill's labour force. This was equally true for chemical pulp as for mechanical pulp. In cases where the wood was actually pulped within the paper-mill, labourers would have been needed to operate the grinders, although not as many as needed in either rag or esparto paper manufacture. As was noted above, the first mechanical wood-pulp plant erected in the States only needed six operators to go into production. Such integrated mills, however, were rarities before the 1890s. Till then a change to mechanical or chemical wood pulp from some other material almost certainly saw a massive reduction in a mill's labour requirements. Non-rag substitutes also reduced employment in the preparatory stage, because, as they were better able to circulate around the engines than

[35] *Ibid.,* p. 15; Hills, *Papermaking,* p. 14; Association of Makers of Esparto Papers, *Esparto,* p. 3.

rags and ropes, they enabled larger engines to be used that could process more raw material with the same amount of labour.[36] An idea of the different labour requirements of mills employing each of the raw materials can be gleaned from information given by Spicer in 1907. According to Spicer, a small rag-using hand-mill needed about twenty-one workers per ton produced; a rag-using machine-mill making high quality paper about fourteen; and a rag-using machine-mill making ordinary grades of paper between four and five. These numbers were considerably above those of the main rag substitutes. In the average esparto-mill two labourers per ton were needed, while in a mill using mechanical wood pulp only one person was needed per *four* tons of paper produced.[37] Esparto thus used half the labour of the rag-mill making ordinary paper, but eight times more than the wood-mill. Paper-mills utilising esparto, straw and wood therefore reduced the amount of labourers needed to process the material. In other words, a changing of raw material from rag to wood or esparto, or even from esparto to wood, would *ceteris paribus* increase the mill's labour productivity. An interesting illustration of this occurred during the Second World War when labour productivity in the paper industry fell by about 17 per cent from its 1939 level. According to Meredith's report on the industry in 1942 for the Nuffield College Social Reconstruction Survey, this was not due to any serious reduction in efficiency during the war, but was the result of 'the greatly increased use of wastepaper and straw which require much more labour in the preliminary stages than the materials for which they have been substituted'. Moreover, the fall in output seriously exaggerated the decline in productivity, since the sectors of the paper trade which declined most in weight in the main were those like newsprint in which 'mechanism counted for most and labour for least'.[38]

To test whether choice of raw material in Britain and America could have influenced employment within the mill enough to have created the gap in productivity would ideally require information on the numbers employed in each country in the preparatory stage of production. Unfortunately, the available data on employment at the industry level are not sufficiently disaggregated to permit such a comparison. However, as was established above, we at least know that with rag paper over half the workforce was involved at this stage in the manufacture of paper, so that anything that altered employment in this department

---

[36] *PTR*, 14 March 1890, p. 2; Spicer, *Paper Trade*, pp. 58, 167.
[37] Spicer, *Paper Trade*, pp. 148–9.
[38] H. Meredith, 'The impact of war on the paper, printing and stationery industries', unpublished paper, September 1942, p.16, held in Nuffield College: Nuffield College Social Reconstruction Survey, C4/119.

would have had a major effect on overall employment in the industry. The unique position of the women in the traditional paper-mill offers us some hope of capturing quantitatively this aspect of the changing pattern of raw material usage.

The role of women in the traditional rag-mill, irrespective of whether it was a hand- or machine-mill, was essentially complementary to the actual process of the manufacture of the paper, which was an exclusively male preserve. Women performed three different tasks within the mill. These were, in order of importance: in the rag room cutting and sorting the rags; in the making up, checking, and packaging of the finished paper; and occasionally as helpers to male labourers in the tending of machines that sized, glazed, rolled, and cut the paper.[39] Since women were thus chiefly engaged in activities that had yet to be successfully mechanised, all improvements in the speed and width of the paper-machines and engines tended to result in a proportionate increase in the demand for women, especially in the rag room.[40] The fact that they were also almost always employed within the rag room, and that the new raw materials tended to reduce women's jobs disproportionately more than any other, enables us to use the percentage of women employed in the industry at different times as a general proxy for what was happening with employment in the preparatory stage. We would expect, therefore, that the less rag used in the manufacture, the less the number of women employed in the establishment. This was certainly the observation of Hutchins writing in 1904: 'The employment of women tends to diminish also in the manufacture of newspaper, where wood pulp has been largely substituted for rags, and rag-sorting and cutting consequently are no longer needed.'[41]

It is also reflected in the comparison of the share of women over thirteen years of age in the workforce of the paper industry of Kent, where there was still a heavy preponderance of rag users, with Lancashire and Yorkshire, where esparto and wood were in much greater use, in 1871. In that year, as expected, 60.2 per cent of the workforce in Kent was female, whereas only 37.3 and 35.5 per cent of employees were woman in Yorkshire and Lancashire.[42]

However, before we go any further, it must be remembered that women were also employed in finishing rooms and that there were other

[39] *Fourth Report on Employment*, pp. 293–4; B. L. Hutchins, 'The employment of women in paper mills', *Economic Journal* 14 (1904), 235–6.
[40] J. A. McGaw, 'Technological change and women's work: mechanization in the Berkshire paper industry, 1820–1855', in M. M. Trescott (ed.), *Dynamos and Virgins Revisited: Women and Technological Change in History* (Metuchen, 1979), pp. 82, 87, 95.
[41] Hutchins, 'Employment', p. 240.
[42] Returns of Factories and Workshops, *PP* LXII (1871), 142–3.

factors that influenced how many female workers were to be employed. In many cases the disappearance of female jobs with the introduction of a new fibre was accompanied by the development of new jobs which could only be filled by men. Wood grinding, for example, was exclusively performed by male workers.[43] Such considerations obviously affect the national percentages given in Table 7.2. With respect to the finishing room women, it is possible to say that, as far as the acceptance of a non-rag substitute in a mill also meant an acceptance of less need for the more intricate finishes to paper and not such a rigorous and painstaking control on quality, new raw materials ought to have reduced female employment in that department as well, and have reinforced the decline in the share of overall female employment. This indeed appears to be what did happen in many finishing rooms.[44] It needs to be reiterated, however, that the percentage of the workforce made up by women remains a blunt instrument with which to dissect this issue, and little meaning should thus be ascribed to it beyond the observation that it is consistent with the thesis that American mills needed fewer female workers because of their greater use of wood and other non-rag fibres.

Table 7.2 illustrates that even from the 1850s the share of women in the workforce of the American paper industry was lower than in Britain. At this stage the difference is not massive, primarily because rag still remained the single most important source of cellulose in both industries. From then, however, a divergence between America and Britain opens up. The gap grows quite gradually in the 1850s, as straw and other materials start to make their presence felt, but after 1860 the divergence accelerates as wood and straw bypass rag as the American industry's main source of fibre. This confirms the observations of contemporaries. In 1887, for example, a group of American paper-makers from Holyoke, who had just spent the summer in Britain visiting eight British mills, remarked that one of the most obvious differences between the two countries was the larger proportion of female labour used in Britain.[45] In Britain the percentage of the workforce made up by women remained relatively high throughout the period, only really beginning its decline in the 1880s when wood became a popular raw material there as well. Even then, the decline was slow and because of the enduring use of esparto and rags was not as dramatic as in the United States. It is interesting to note that this slow decline in female employment in Britain continued in the interwar years, so that by 1939 less than a quarter of the industry's workforce were women.[46] Spicer

[43] Hutchins, 'Employment', p. 240.   [44] McGaw, *Most Wonderful Machine*, p. 286.
[45] *PMC*, 10 December 1887, p. 432.
[46] Meredith, 'Impact', p. 13 in Social Reconstruction Survey, C4/119.

Table 7.2 *Share of women in the workforce in the UK and US paper industries, 1850/1–1905/7*

Year	UK	US
1850/1	48.6	43.5
1860/1	47.2	40.3
1870/1	43.7	34.4
1880/1	48.8	32.1
1890/1	43.3	21.8
1900/1	40.0	16.0
1905/7	32.6	13.5

Sources: Spicer, *Paper Trade*, p. 146; *Seventh to Thirteenth Census of Manufactures* (1850–1905).

attributes some of the gradual decline in women's share to the Factory Acts Extension Act of 1867 which limited female work in paper-mills to daylight hours.[47] Hutchins, who investigated the effects of the Factory Acts on female employment for the Committee of the British Association in 1904, however, could find no evidence that very large numbers of women were affected by this legislation or that the provisions caused any serious inconvenience. This stemmed from the fact that since the work of men and women was essentially distinct, there was no need to have the men and the women working side-by-side at the same time. Moreover, as it was generally held by papermakers that rag-cutting and sorting needed bright sunlight and could not be done by artificial light, the restricting of the hours of women's work to between 6a.m. and 6p.m. did not alter existing hours which even prior to 1867 were usually from 8a.m. to 5.30p.m.[48] The increased share of women in the workforce in the 1880s confirms this conclusion.

The linking of the rapid decline in female employment to the advent of wood is also supported by the experiences of individual firms in Britain. In Table 7.3 the share of all female employees in Sommerville and Co.'s mill in Dalmore on the Esk are presented. This was a mill that specialised in fine printing and engine-sized writing papers. In 1871 it made these papers from rag and esparto. By the 1880s, however, it had started to experiment with wood pulp, steadily introducing larger percentages of it into its paper stock, so that by the early 1890s it was a major source of fibre for the firm. The decline in the proportion of employees who were female corresponds with the introduction of wood pulp into the mill.[49]

Choice of raw material, therefore, had consequences for the amount

---

[47] Spicer, *Paper Trade*, p. 147.   [48] Hutchins, 'Employment', pp. 237–8.
[49] Watson, *Last Mill*, pp. 27, 145.

Table 7.3 *Share of female workers in the entire workforce at the Dalmore Mill, 1871–1906*

Year	Percentage
1871	54.6
1881	59.6
1883	50.0
1889	42.3
1896	32.9
1900	33.3
1906	24.9

Source: Watson, *Last Mill*, p. 145.

of labour employed, and through this medium, could have direct influence on the level of labour productivity achieved. This is what the papermaker, W. Nuttal of Cooke and Nuttal Limited, certainly believed: 'The large increase of paper made in this country has not led to a proportionate increase in labour, because the pulp, made of wood and largely used now, is a semi-prepared article, whereas we formerly had to produce pulp from rags, etc. ourselves.' Nuttal, however, acknowledged that it was not the only factor at play: 'This is of course not altogether due to the difference in methods of production and the substitution of wood pulp for rags, but is also the result of economies of labour, [and] quicker running machines.'[50]

Labour-saving machinery

Clearly the introduction of machinery also affected labour productivity. Along with its greater use of non-rag fibres, the greater proclivity of American paper-mills to utilise more labour-saving machinery was the other major component that contributed to the labour productivity lead that that country held over Britain. References by contemporaries to this fact from the middle of the nineteenth century are manifold. Dyson, for example, strongly believed that 'mechanical appliances for labour-saving are also more largely used than is the case in England, making it much easier for the workmen generally'.[51] British commentators marvelled in particular at the achievements their American counterparts had made in streamlining the flow of materials through the mill with the aid of mechanical devices that economised on expensive labour. As early as

---

[50] Tariff Collection, TC7 28/2, p. E(1).
[51] Dyson, 'Paper Trade', p. 215. Other examples are *PMC*, 10 March 1903, p. 98; 10 July 1901, p. 306; Reed's comments in Tariff Collection, TC3 1/88, p. 7.

1862 we find the British trade press lauding the progressiveness of the Ivanhoe Mill in Paterson, New Jersey: 'The buildings are admirably arranged, and whilst every department ... is complete within itself, the whole is so connected as to be perfectly progressive; thus economising on labour, and preserving the orderly arrangements and cleanliness of the whole establishment.'[52] Thirty-two years later the same feature still impressed visiting British papermakers who attributed America's productivity leadership at least partially to this factor: 'In the paper mills at Holyoke they go in for machinery to save labour greatly and they have some very neat appliances for conveying their rags and stuff from one department to the other.'[53]

The labour-saving machinery used in America touched on all aspects of the production process from rag-cutters which replaced women in the sorting room to self-feeding calenders that saved on labour in the finishing room. An example of a machine popular in American but rare in British mills was that which aided in the replacement of old paper-machine wires. The changing of the wire on the paper-machine was a labour intensive and time-consuming process that had to be done fortnightly. Without any help from machinery it usually took the full complement of the machine room a good six gruelling hours to remove an old wire and fit a new one. In America the same procedure by the second half of the century had been mechanised by appliances that lifted the breast roll, couchers, and press rolls, so that a handful of men could do the job in a mere eighty minutes.[54]

Comparable data that would permit us to plot the long-term progress of capital intensity in the two countries from the mid-century are unavailable. In chapter 3 paper-machine numbers were used as a measure of capital, but since these were not calculated in the United States until very late in the nineteenth century, this avenue is not open to us. Another commonly used measure for capital is the energy requirement of the industry. Whilst this time American censuses do provide such figures, there was no regular publication of energy use in Britain. We are not, however, left without any idea at all. The Returns of Factories and Workshops in 1871 do give us enough information about horsepower consumption in the British paper industry to allow at least a comparison with the United States in that year to be made.

On the assumption that the more machinery used, the greater the energy needed, this snapshot confirms the view of contemporaries that labourers in the United States had access to more machinery than their counterparts in Britain by a margin of more than two-to-one. It also

[52] *PTR*, December 1862, p. 29.   [53] *PMC*, 11 June 1894, p. 278.
[54] *PMC*, 10 March 1903, pp. 97, 103.

Table 7.4 *Energy consumption in the UK and US paper industries, 1870/1* (*horsepower*)

	Steam power	Water power	Per ton of output	Per man year
US	11,922	41,684	0.16	2.85
UK	26,703	7,660	0.16	1.30

Sources: Returns of Factories and Workshop, p. 218; *Ninth Census of Manufactures* (1870), p. 463.

suggests that the productivity of capital in both countries was similar. Although it is hazardous to make strong claims on the basis of one observation, if we read the horsepower per ton of output figures literally, it implies that American paper manufacturing in 1870 was apparently no more wasteful of energy than British. The same cannot be said for its use of papermaking materials. Employing the estimates of raw material usage in both countries in 1880 that were calculated earlier, it can be shown that on average American paper manufacturers utilised over twice as much per ton of paper and board produced than the British: approximately 4.1 to 1.7 tons. Given that these figures are at best only rough estimates and that British papermakers preferred materials such as rag which tended to have a higher percentage of cellulose per unit of weight than most other papermaking materials, the scale of the gap between the two industries suggested by the estimates may be somewhat overstated. Nevertheless, there is little reason to doubt its existence to some degree, as America's profligate use, at least from the European perspective, of raw materials, especially in the wood and forest-related industries, is one of the hallmarks of nineteenth-century American industrial practice.

What explains America's greater use of labour-saving machinery? Like its choice of raw material after the Civil War, the explanation appears to lie in relative factor prices. It was seen earlier how wages for all occupations in the American paper-mill were up to double those prevailing in Britain.[55] Although similar information on the price of

[55] While it is true that the American paper industry's greater capital intensity may have contributed to its relatively high wage rates by raising productivity, the fact that this state of affairs was a feature of virtually all American industry in the nineteenth century implies that the causation was more likely to have run the other way. In other words, American wages in papermaking were higher than British wages because of specific features of the American environment (e.g. the frontier) rather than the greater capital intensity of its papermaking. The introduction of machinery may have put additional upward pressure on wages, but this would only have had the effect of causing further substitutions of capital for labour to have been made.

machinery in both countries is not available, it is noteworthy that British papermakers visiting America expressed surprise at the low prices there; a surprise that presumably stemmed from the fact that in the 1880s paper-machine making had still only had a relatively short existence on that side of the Atlantic.[56] Its development, however, had been rapid, so that even by the mid-century American paper-machine manufacturers had not only well and truly established themselves, but had already become an experienced and innovative force in the paper trade, producing a wide variety of quality and affordable products. Long gone were the days when local papermakers out of necessity were compelled to rely on imported foreign parts, felts, wires, or dandy rolls to keep their paper-machines running.[57] More than anything else, however, it was the declining relative price of papermaking machinery in the American industry in the second half of the century that best illustrated just how far its machine-makers had come of age.

Since the 1830s clusters of specialist paper-machine manufacturers had set themselves up in all of the major papermaking regions of the country. These firms quickly became both technologically and commercially more sophisticated, in many cases offering by the post-bellum period, for example, trade-in deals on old machinery to their customers. In addition to these, many machine shops originally interested in other types of machinery, usually textile and agricultural, also branched out into catering for the needs of local papermakers. In Berkshire County, Massachusetts, alone there were six such non-specialist machine shops and foundries making parts, tools, and machinery for papermaking in 1860; twenty-five years later, that number had more than doubled. This rise and clustering of machine shops in the major papermaking regions of the country since the 1840s progressively lowered the cost of machinery to papermakers. This was primarily due to the experience and expertise that the highly competitive environment had nurtured, but lower transportation costs, as well as the quick instalments and repairs, and hence reduced downtime for papermakers that they could guarantee, were also factors.[58] A further consideration that augmented the cheapness of machinery in America was the not uncommon practice of neighbouring mills sharing equipment that needed to be used only infrequently. In Berkshire County, the papermaker Byron Weston combined his resources with the adjoining Bartlett and Cutting partnership to purchase a calendar lathe. As it became more common for firms to own more than one mill, the incentive to adopt such practices grew

[56] For example see *PMC*, 11 February 1889, p. 52.
[57] Keir, *Manufacturing*, p. 482; Weeks, *History*, pp. 182–3.
[58] McGaw, *Most Wonderful Machine*, pp. 168–76.

ever stronger. From 1875, for example, the four mills operated by the Smith Paper Company shared not only the machines needed for making quick repairs, but even a machine shop.[59]

The other crucial factor that acted to lower the expenses of using machinery in America was the abundance of natural resources. Certainly British papermakers looked upon America's endowment of fast running rivers and ready supply of cheap energy with noticeable envy.[60] American manufacturers too seemed to have been well aware that few other nations were so blessed with water power.[61] If the figures for energy use in both countries supplied in Figure 7.4 are decomposed between its source, steam or water, some justification for the belief of contemporaries is possible. In America in 1871 77.8 per cent of the total energy used was generated by water, while in Britain the figure was only 22.3 per cent. As water was a relatively easy resource to harness, this abundance of easily accessible energy not only enabled American producers to run their machines more cheaply, but given their large wage bills, also provided them with strong economic incentives to introduce more machinery in their mills. In a similar manner America's greater access to cheap and ready supplies of cellulose such as wood and straw made the introduction of machinery that saved on labour, but was extremely uneconomical in terms of raw material wastage, a worthwhile venture. In Britain, the balance of factor prices arising from its own unique resource endowment ensured that such an option could not be viably pursued there.

Thus, American papermakers opted for more labour-saving devices because it made good economic sense for them to do so. As McGaw explains, at least with respect to the American industry's early mechanisation: 'The initial cost of newly developed machines, though substantially higher than that of hand tools, totalled far less than the wages saved, so that the rapid acceptance of novel machines made good economic sense.'[62]

It made sense simply because the cheapness of the machinery in conjunction with the availability of inexpensive energy and materials provided adequate incentives to papermakers to substitute capital for relatively expensive labour. In this regard the paper industry appears as

---

[59] Ibid., pp. 172–3.
[60] For example Dyson, 'Paper trade', p. 216; Paper Maker and British Paper Trade Journal (hereafter PMBPTJ), November 1902, p. 49.
[61] See, for example, Pulp and Paper, vol. II, p. 433; D. L. Boese, Papermakers: The Blandin Paper Company and Grand Rapids, Minnesota (Grand Rapids, 1984), p. 55; and E. Amigo, M. Neuffer, E. R. Maunder, Beyond the Adirondacks: The Story of St. Regis Paper Company (Westport, 1980), p. 4.
[62] McGaw, Most Wonderful Machine, p. 179.

one of those skilled-manufacturing industries in America that exhibited greater capital intensity than was experienced in Britain.[63]

Summary and conclusions

Between 1860 and 1890 two characteristics of the American industry kept its labour productivity twice that of the British. One of these features was the different choice of raw materials that each of the nations made in this period. In the United States this choice reflected both its unique history and economic environment. The intolerable shortage of rag that the Civil War imposed on the American paper industry made the continued use of that material there impossible, prompting a widespread experimentation with, and introduction of, raw materials such as wood that, as in Britain, had not been previously taken seriously. As a consequence, however, the American industry attained an experience and expertise with a number of non-rag substitutes that had hitherto been unavailable in either Britain or America. In particular, the technical weaknesses that had marred earlier attempts to use wood were progressively removed in this period, so that by the end of the 1860s when a semblance of normalcy had returned to the industry, American papermakers found that the choice of raw materials that now confronted them had significantly broadened. No longer need rag be their automatic choice, simply because there were no other alternatives available to them. Instead, American papermakers were now free to choose from a wide range of materials. The factor and resource costs that prevailed in the second half of the century ensured that the low labour requirements and relative cheapness of wood pulp quickly made it the favourite of the American papermaker.

In moving away from rag and selecting materials like wood that needed fewer workers, American manufacturers were also consciously altering work practices in their mills. The transition from rag paper to wood paper in particular brought about dramatic transformations in manning requirements and wrought a major change in the role of women in papermaking. But such changes not only altered the composition of the paper industry's workforce, but also significantly affected the nature of work carried out, for prior to the introduction of these new sources over half of all workers employed in paper-mills were engaged in the preparation of the rags. The number of such workers needed with non-rag substitutes was significantly less. As a result, the

[63] J. James and J. S. Skinner, 'The resolution of the labor-scarcity paradox', *Journal of Economic History* 45 (1985), 513–40; N. Rosenberg, (ed.), *The American System of Manufacturing* (Edinburgh, 1969).

more rapid diffusion of these new raw materials in the United States enabled manufacturers there to produce the same quantity of paper with less labour than in Britain and than was the case before.

In conjunction with the advent of wood and straw paper, American papermakers also employed labour-saving machinery more freely. This was very much a response to the relatively high cost of labour as well as the bountiful endowment of easily accessible water power and paper-making materials in America. Clearly resource endowment, both in terms of the type of resources on offer as well as their availability, played a vital and influential role in the distinctive developments of the British and American paper industries in the third quarter of the nineteenth century. The influences of the environment on the industry were diverse. Thus, raw materials were important not just as cheap inputs into the production process but, as we saw with regard to the choice of raw material, in determining both work and manning practices in the industry; factors which had importance for labour productivity as well. As such then, in as far as Britain's delayed adoption of wood pulp was rational and the American industry's greater capital intensity and choice of raw materials also reflected its unique history and resource endowment, it is hard for one to conclude that the emergence of the labour productivity gap between 1860 and 1890 had its origins in the widespread failure of British papermaking.

8 Technological divergence: the Anglo-American gap, 1890–1913

In chapter 7 it was argued that, up until 1890, the differential in the level of labour productivity experienced by Britain and the United States since the mid-century could primarily be attributed to the effects on the employment of labour within the mill, that the raw materials and the amount of labour-saving machinery used in each nation had had. From 1890, however, it becomes more difficult to explain the persistence of the productivity gap purely in terms of factors that influenced employment practices. As we saw from the data on machine speeds and capacities, as well as total productivity, given in chapters 3 and 5, it is from this date that a technological lead starts to appear in the United States. But if America was pulling ahead of the British technologically, the question must be asked why this was possible?

Differences in the technological paths adopted by each nation may provide an explanation, especially since in earlier years the American industry had made use of different types of machinery. This notwithstanding, however, the fact remains that the basic technique for making paper in the last half of the nineteenth century was for all intents and purposes the same in these countries. Moreover, even if we assume this not to be the case, it was the Cylinder machine – that was slightly more popular in the States – and not the Fourdrinier that had less potential for improvement and speeding up. Differences in the type of paper-machine used cannot account for America's technological lead in the quarter century before the First World War.

Another possible explanation is that unions and work practices in Britain placed restrictions on the speed and/or the number of hours that a paper-machine could be operated. However, as discussed in chapter 6, there is no evidence at all that unions or customary work practices were important factors in British mills. Trade unions in the British paper industry were small, divided, and weak, and incapable of even attempting to impose such restrictions. Ironically, it was in America that the labour movement in the paper trade had more potential. In any case, 1890 to 1913, far from being a period in which machine speeds and

Table 8.1 *Structure of UK and US paper production, 1905/7 (in percentages)*

	UK	US
Paper for printing, newsprint, and posters	52.4	46.9
Packing and wrapping paper	21.7	21.2
Pasteboard, cardboard, and millboard	6.2	17.1
Envelopes and fine writing and drawing paper	13.6	4.8
Other types of paper and board	6.1	10.0

Sources: Census of Production (1907), table 1, p. 624; *Thirteenth Census of Manufactures* (1905), table 20, p. 757.

work practices stagnated, was in fact a time when mills in both countries were driven by a frenzy to increase operating speeds and time. In light of British papermakers' well-known contempt for the quality of American paper, a more likely explanation is that British manufacturers concentrated on a higher quality of paper, for which speed was not only unimportant, but positively deleterious to the final product. Looking at Table 8.1, however, this too becomes improbable. In fact, perhaps contrary to expectations, a larger share of British production in 1907 was devoted to newsprint and printing – where speed was paramount for survival – than of American. Whilst a greater share of British production was also made up by the finer grades of paper than was true in America, in terms of the sectors in which there was good reason to maximise speed – printing, packing, and boards – there is very little difference in the shares of total production in both places. What is more, these sectors together were no insignificant part of the industry; representing over 80 per cent of all paper and board produced in Britain at that time. Moreover, there is no reason to believe that the structure of the British industry was any different in 1890. Although we do not have the same information for that period, it is possible to say that virtually the same percentage of firms made newsprint, printing and fine hand-made paper in 1890 as in 1907.[1] This, of course, does not make any allowance for the scale of production in each mill and so does not truly reflect shares of production; it does, however, because of the greater scale of printing and newsprint establishments, imply that the structure probably did not change very much between 1890 and 1907. The sectoral composition of the British paper industry does not provide an explanation for the differing technological performance.

[1] In 1890 33.8 per cent of British mills made newsprint and printing paper and 6.5 per cent hand made paper. For 1907 the respective figures were 36.8 and 6.4 per cent. Calculated from the annual publication *Paper Mill Directory*.

Another way to approach this question is to look for incentives, present in one country and absent in the other, that could have induced a greater degree of innovative effort and success in one of the countries. Such an approach, however, sheds little light here because between 1890 and 1914 both the American and British paper industries were bombarded by the same intense incentives to speed up production. In any case, at any time there always exists a plethora of incentives for a producer to innovate; how one exactly distinguishes between those incentives that are acted upon and those that are not is unclear. Moreover, positing the notion that a particular incentive leads to technological change leaves those instances where it does not totally unexplained. By implication such unexplained instances also strongly suggest that there must be other factors at work as well in those cases where successful innovation does seem to follow from the presence of a certain incentive. In other words, to understand the process of technological change as it actually takes place, rather than in an abstract world of perfect rationality and immediate response, one must be able to explain why particular innovations occur in particular firms at a particular time and not elsewhere at a different time and in different contexts. To do this requires one first to understand the nature of technological change in the industry, and then to examine the effects of setting and context on this type of technological change.

It will be recalled from chapter 2, where the process of technological change in the industry was examined, that innovation in late nineteenth-century paper machinery took the form of a gradual accumulation of technological knowledge, that added to, rather than superseded, existing technology. Such change was by nature cumulative, incremental, and in many cases firm-specific. These basic characteristics are, of course, familiar to students of other continuous production technologies. What is less familiar and certainly little understood are the forces that influence the rate at which technological accumulation occurs in particular firms and industries. In chapter 2 it was argued that with this brand of technological change, there are essentially two mechanisms through which firms can progress technologically.

The first revolves around the economies of practice that follow from everyday experience in production. By implementing measures, either intentionally or otherwise, that support this gradual process, an industry (or firm) can enhance its possibilities for internally generated techno-logical change. These greater possibilities are achieved by providing those in the firm or industry with better opportunities and incentives to learn from experience and, on the basis of this learning, to suggest and ultimately implement improvements that either raise best practices or extend existing technology. In the pursuit of such improvements the potential of the technological system as well as *inter alia* the industry's attitudes and policies on collective action, business organisation, training, and labour relations are of vital importance.

The second mechanism, although not necessarily mutually exclusive from the first, offers firms so wishing the opportunity to bypass the first. Instead of generating their own technological initiatives, manufacturers could simply purchase innovations and new equipment from specialist producers and other firms whenever they become available. Reliance on such a strategy has its drawbacks, not the least in that it can be extremely expensive. However, in conjunction with the first mechanism it represents a potent strategy that could keep the process of technological change going even when internally generated innovation has stalled.

In the remainder of this chapter the technological lead in papermaking acquired by the American industry from the 1890s is viewed through the filter of these mechanisms to see if any systematic differences between Britain and the United States can be identified which might explain the different rates of technological accumulation experienced. This approach to technological creativity has advantages in that it is firmly based on the actual process of technological change in the industry in this period as it occurred, and is thus less dependent on *recherche* and often unrealistic abstractions of the other approaches. In the final section of the chapter a summary of the last four chapters' findings and their implications is given.

New machines

As with most industries, one of the most common ways technological change was transmitted in the nineteenth-century paper industry was through the purchase of new machinery. In this process American paper manufacturers seem to have been more adept than their British counterparts. Indeed, one thing that British visitors to the United States frequently commented on was the ease with which American firms, even the older mills, replaced their equipment. On returning to Britain, Dyson, for example, remarked: 'In England, we have very good and expensive machinery, but in America, when the machine has paid for itself, they put in a new one. Our machines last too long, and the employers hesitate to change them for something better.'[2]

---

[2]   *PMBPTJ*, November 1902, p. 49. Similar remarks can be found in J. Dunbar, *Notes on the Manufacture of Wood Pulp and Wood-pulp Papers* (Leith, 1894), p. 55; *PMC*, 11 September 1893, p. 382; 11 May 1903, p. 172; 10 March 1903, p. 103; Tariff Collection, TC3 1/85, p. 27.

Little explanation for this fact can be found in the activities of British paper-machine makers who appear to have operated similarly to their American counterparts. They made trips to the Continent and United States to view new developments, take back ideas, and negotiate licences for the more successful foreign patents. The records of various British firms confirm that through correspondence and frequent visits there was indeed an effective liaison between papermaker and paper-machinist.[3] The paper-machine maker, Andrew Masson of Masson, Scott and Co., could thus honestly claim after his trip to the United States that:

I came over to see what new wrinkles, if any, I could pick up. Your machinery is very good; but not, I think, any better than ours. There are some points of difference, chiefly in smaller details, between the English and American papermaking machines ... I do not think that our machine makers need fear competition from their side, nor is it probable that we can compete with you ...[4]

Paper-machine makers testifying before the Tariff Commission in the next decade echoed these views, in addition asserting that if paper-machines on average were less advanced in Britain, this was due to the papermaker and not the machinist.[5]

One way to get some idea of the differing replacement and acquisition rates in the respective countries and test the widely held perceptions of contemporaries about these rates is to look at the rate of change over time in the average width of paper-machines in both industries. This is a good proxy because width is one feature of the paper-machine that could not be altered by improvements within the mill. That is, to get wider machinery a mill could not simply weld on a few more inches of steel to existing ones, but would actually have to buy a new one from a machine maker who had already solved all the detailed and complicated technical problems associated with making machines that little bit wider. As such, for each new and wider machine added to the industry's stock, the average width of a machine in that stock should rise. Moreover, the more frequently new machines are installed, the faster the increase in this average. These averages for Britain and America for selected dates between 1868 and 1914 are given in Table 8.2.

An interesting feature of Table 8.2 is that the annual average compound rate of growth of American widths is almost double that of the British over the whole period: 1.01 per cent as compared to 0.54. These rates remain fairly constant. Breaking the US data into three periods of approximately fifteen years (1868–85, 1885–1900, 1900–14)

[3] PMC, 10 December 1891, p. 470; Shorter, *Paper-making*, p. 138.
[4] PMC, 10 September 1896, p. 440. Also PMC, 10 December 1891, p. 470; 10 July 1899 p. 302; 10 December 1889, p. 453; and 10 March 1904, p. 95.
[5] Tariff Commission, *Engineering*, paras 132, 1255, 1260.

Table 8.2 *Average width of paper-machines in the UK and US, 1868/9–1914*

Year	UK	US
1868/9	64.7	58.9
1885/6	70.0	62.9
1888/9	71.7	64.6
1893/4	73.2	70.9
1897	75.1	74.5
1900	77.3	75.0
1903	77.7	77.4
1906	78.5	79.8
1909	80.5	84.5
1912	81.4	87.5
1914	82.8	88.4

Sources: *Paper Mill Directory* (1869–1914); *Paper Mill Directory of the World* (1897–1914); *Lockwood Directory of Paper Manufactures in the United States and Canada* (1885/6, 1888/9, 1893/4); J. A. Murphy, *1868 List of Paper Manufacturers*.

yields growth rates of 1.00, 1.01, and 1.01 per cent. The same constancy is found in British growth rates in these periods. The implication of Table 8.2 is that the belief of contemporaries about American replacement rates was probably not unfounded. Their failure to notice it prior to the 1890s was presumably related to the fact that it was only at the end of the century that British papermakers were beginning to grow seriously concerned with the relative progress of their industry. After all, before that date American paper technology on average appears to have been no better than British and may have even lagged behind.

What accounts for the American papermaker's more frequent replacement of machinery? One of the most important features of the second half of the nineteenth century was the vast growth in demand for paper. This was most noticeable with newsprint. In the four decades between 1860 and 1900 total newspaper circulation in the United States grew at an average annual compound rate of 5.5 per cent from 13,663,409 to 114,299,334 tons.[6] The growth in demand, however, was not only due to the spread of literacy and cheap wood-pulp paper, which had sparked the proliferation of newspapers with ever larger circulations, but was also related to the increasing number of uses that were now being found for paper products. In these vast changes in consumption habits some have seen a paper revolution.[7] In this half century mass-produced paper bags,

[6] D. C. Smith, 'Wood pulp and newspapers, 1867–1900', *Business History Review* 38 (1964), 345.
[7] Smith, *History*, p. 139; Hunter, *Papermaking*, pp. 385–9.

containers, and other forms of packaging emerged, as did the popularisation of the postcard and cheap fiction. Remarkable new technologies like photography and telegraphy arose requiring new varieties of paper, and older ones developed in new directions. In food retailing the general store, which had previously bought items in large bulky containers, dividing and repackaging the goods themselves into more convenient and saleable amounts in the shop, gave way to the modern, self-service store with its wide range of brands and goods each individually prepackaged by the manufacturer for convenience and ease of identification.[8] In the printing trade new special surfaces, such as half tone for reproducing photographic images and coloured stock, were called for and the paper industry responded. Paper also found more uncommon uses, for example, as a building material for boats, coffins, and even torpedos. Paper shirt fronts and handkerchiefs also became popular fashion accessories, while a series of clean and disposable personal hygiene items made from paper emerged and rendered everyday life more salubrious. In 1871 the toilet roll appeared, and by the end of the century its use was universal in Western societies.[9]

Population growth in America of the second half of the nineteenth century was also relatively rapid. In 1860 America's population was already one-and-a-half times larger than Britain's, but by the turn of the century it had grown to close on two-and-a-half times the British figure.[10] Combining the enormous size and rate of growth of the American population with the explosion in the uses of paper and board after 1860, the intrinsic opportunities available to the American papermaker to exploit are obvious. Protected by tariff walls and endowed with an abundant supply of ready capital, especially in the last quarter of the century, American manufacturers had this enormous and growing home market virtually to themselves.

Moreover, the spread of the railroad after the Civil War removed the geographical constraints on this market. In 1860 with 30,000 miles of track in operation America's rail network was already without doubt the largest in the world. Yet in the aftermath of the Civil War another period of enormous expansion, unprecedented in its magnitude, was embarked upon, which steadily brought even the most remote regions of the

---

[8] Cereal and biscuit manufacturers pioneered this direction. H. T. Bettendorf, *Paperboard and Paperboard Containers: A History* (Camden, 1946), p. 27; Karges, 'David Clark Everest', p. 322.

[9] H. E. Wright, *Three Hundred Years of American Papermaking* (Washington, 1991), p. 18; Hunter, *Papermaking*, p. 570 and *passim*.

[10] US population was 31,443,000 in 1860 and 75,994,000 in 1900. The British figures were 20,646,000 and 36,686,000. Mitchell, *Historical Statistics*, pp. 11–13; Smith, 'Wood pulp', p. 345.

country into contact with the large markets of the Eastern seaboard and the old Northwest. In 1860, for example, Wisconsin had only 891 miles of railroad, but by 1890 this figure had swollen by 626 per cent to 5,583 miles: an annual compound growth rate of 6.3 per cent. Such expansion allowed individual manufacturers, formerly restricted to producing for local and nearby markets, to tap for the first time into the large, standardised market.[11] And many did, taking full advantage of the opportunity to expand, specialise, and later integrate to a degree hitherto unknown in the trade. New firms were also enticed into the industry, with the number of establishments reported in censuses increasing from 649 in 1889 to 777 in 1909.[12] Of course, the same trends in consumption and production (except the number of mills which fell) were felt in Britain where per capita consumption of paper was as high, if not higher, than in America, but given the limitations and size of the home and traditional export markets, the magnitude of their effects could not take on similar proportions.

Illustrative of this difference were the actual quantities of paper consumed and produced in both countries. As Table 8.3 shows, a big gap in the relative consumption of the two nations appeared most noticeably from the 1890s, and by the First World War America's consumption of paper had grown to over three times that of Britain. Between 1879 and 1912/14 the average annual increase in consumption in America stood at 7.2 per cent and in Britain 4.6 per cent. A similar picture is given by the relative production figures in Table 8.4. Over the same period British paper production grew on average at 4.9 per cent per annum compared to the much faster American rate of 7.5 per cent: growth rates that had been fairly consistently maintained since the mid-century. This faster rate in America had important consequences for the replacement of machinery there. In theory, *ceteris paribus*, if one industry grows faster than another, it will have on average a younger capital stock, and if, in turn, technological change is largely embodied in new capital equipment, this newer capital will be more efficient, granting the rapidly growing industry a lower cost structure. Assuming that investment and output growth and depreciation stay at constant exponential rates, the

---

[11] M. L. Branch, 'The paper industry in the Lake Regions, 1834–1947', Ph.D. thesis, University of Wisconsin (1954), p. 27; W. B. Wheelwright and S. Kean, *The Lengthened Shadow of One Man* (Fitchburg, 1957), p. 14; J. M. McPherson, *Battle Cry of Freedom: The Civil War Era* (New York, 1988), p. 12; A. D. Chandler, 'Development of modern management structure in the US and UK', in L. Hannah (ed.), *Management Strategy and Business Development* (London, 1976), p. 47; E. Rothbarth, 'Cause of the superior efficiency of the USA industry as compared with British industry', *Economic Journal* 56 (1946), 386.

[12] *Fourteenth Census of Manufactures* (1909), table 2, p. 750.

Table 8.3 *Comparative consumption of paper and board in the US and UK, 1859–1912/14*

Year	US/UK
1879	119.2
1889	125.6
1899	167.5
1907/9	289.6
1912/14	322.0

Notes: British figures for 1879, 1889, and 1899 are estimated by interpolation from 1875, 1885, 1895 and 1907 to make them compatible with US data.
Sources: Wray, *Study*, p. 219; Frickey, *Production*, p. 16.

Table 8.4 *Production of paper and board in the UK and US, 1860–1912/14 (long tons)*

Year	UK	US
1860	111,905	113,400
1870	169,023	344,600
1880	229,638	403,600
1890	411,483	834,800
1900	699,404	1,935,700
1912/14	1,085,243	4,705,400

Sources: Hoffmann, *British Industry*, T54; Frickey, *Production*, p. 16.

average age of capital in industry can be calculated as the reciprocal of the sum of these two rates.[13] Thus, if the rate of depreciation in paper machinery was 5 per cent – the rate allowed by the US Treasury Department[14] – then in 1913 American machines were on average eight years and British machines just over 10 years (10.1) old. This seems fairly close to reality, for at the time of its establishment over two-thirds of the International Paper Company's machines were less than a decade old, whereas most British papermakers seemed to have operated machines of ten or more years old.[15]

---

[13] R. M. Solow, 'Investment and technical progress', in K. J. Arrow, S. Karlin and P. Suppes (eds.), *Mathematical Methods in the Social Sciences 1959* (Stanford, 1960), pp. 89–104. This technique is used for the steel industry in Temin, 'Relative decline', p. 149.

[14] Stevenson, *Background*, p. 89.

[15] Vandermeulen, 'Technological change', p. 132. If repaired and upgraded machines could last for decades. Number 3 machine bought by Guard Bridge Paper Company Limited in 1887, for example, was still running in 1912 when it was electrified. Weatherill, *One Hundred Years*, p. 25.

To match the American growth rate, and hence its rate of embodied technological change, the British industry would have needed to have had an export growth rate of 12.5 per cent per annum between 1881 and 1912, instead of the 5.9 per cent that was actually achieved. That is the equivalent of requiring exports' share of total British production in 1912 to increase from its actual share of 15.3 to 49.6 per cent. Even if it is assumed that the British industry was able to secure for itself through protection the entire domestic market for paper, the industry would have still needed to have had an annual average compound rate of export growth between 1881 and 1912 of 11.5 per cent just to keep up with the Americans. Given that the other major markets of the world were effectively blocked from the British manufacturer by tariff walls and that even its traditional markets in the empire were increasingly being penetrated by American and continental paper, or like Canada and Australia were in fact developing their own industry, the likelihood that the magnitude needed could be achieved was highly improbable.

But what part did America's faster rate of demand play in the emerging technological gap in the decades immediately before the First World War? To a large extent the answer to this question depends on the rate of embodied technological change experienced. In the paper industry the best way to get an idea of this is to consider machine speeds. Given that the increase in both maximum and average machine speeds never exceeded 6 per cent per annum, our calculated 2.1 year difference in the mean age of paper-machines at the very most could have accounted for a productivity differential between American and British machines of 12.6 per cent. This is certainly no trivial amount, but it still leaves most of the 36.2 per cent differential in average machine speeds and the 48.1 per cent differential in average machine capacities in 1912–14 unexplained. The estimated figure also assumes that the best-practice technology in each country was equivalent. However, if the state of papermaking technology in reality had differed, this fact would also have had an effect on the degree of technological sophistication found in average machinery. For example, if it could be shown that American papermakers were pushing back the technological frontiers of their trade faster than the British, this factor would have accelerated even further the rate of embodied technological change experienced. Looking back to Table 5.5, which showed the speeds of the *fastest* recorded machines in operation, a reasonable proxy for the state of best-practice technology at any date, we discover that from the 1890s there are grounds to believe that this was indeed the case. To what can this spurt of technological ingenuity be attributed? One obvious explanation is that the demand which encouraged American papermakers to replace and to acquire machinery more frequently also spurred them on to innovate and invent more. Given the facts that American output constantly grew at between 2.5 and 3 per cent faster than British output at all times from 1860, but maximum speeds in both countries only began to diverge from the 1890s, this explanation appears unsatisfactory.

An alternative approach to such questions is to look to the other mechanism through which papermaking firms progress technologically, the economies of practice, to provide answers. After all, not all technological change was embodied in new machinery. It will be recalled from chapter 2 that attaining an understanding of this source of technological change requires consideration of those factors that affect the industry and the firm's opportunity, ability, and incentive to learn from production. Such an analysis is conducted in the following sections where four systematic differences that explain the divergent technological performance of the two industries from 1890 are examined.

Trade associations

One area where the British and American industries differed was in their attitudes to trade organisations and industry-wide co-operation. Unlike the British industry, which was highly competitive in structure and outlook and officially abhorred combination, American manufacturers from an early date showed an interest in the benefits of self-regulation and co-operation. In part this was due to a belief that such behaviour would enhance profits and facilitate technological change, but there was also a feeling of higher duty amongst many mill-owners to promote formally and informally the industrial development of the community and industry that had nurtured them.[16]

The first recorded meeting of papermakers in America took place in May 1819, at which a petition was drawn up requesting Congress to increase import duties. Thirty years later the issue of protection once again provoked the calling of a convention, this time to compile a memorial for Congress to protest against the reduction of duties on paper. For this purpose statistics on the amount of capital invested, the number of labourers employed, and the number of mills operating in the industry were collected. Although no permanent organisations resulted from either of these meetings, they were indicative of an early willingness amongst papermakers to co-operate on matters of interest to all in the trade.[17] It was not until February 1861, when twenty-one out of the

[16] McGaw, *Most Wonderful Machine*, p. 380.
[17] Carlson, 'Associations', pp. 34–5, 293.

thirty-six manufacturers of fine writing paper met in Pittsfield, Massachusetts, that the industry had its first proper trade organisation, the Writing Paper Manufacturers Association. This was essentially devised as a protective organisation, though it was also hoped that it could be used as a forum for discussing issues of relevance to its members. After its first meeting plans were made to implement voluntary restrictions on output to counter a perceived glut in the market, but these had to be shelved when the Civil War broke out later that year.[18] During the war the Paper Manufacturers Association of the United States, the industry's first *national* trade organisation, was formed to deal with the exigencies the war had thrown up. Little is known about this association and its activities, except that before its demise sometime between 1867 and 1870 it had tried to establish a statistical bureau for the benefit of its members and the government.[19] By 1873 the Writing Paper Manufacturers Association, still comprising three-quarters of the mills in the sector, had reconstituted itself and once again begun discussing measures to remedy the adverse economic situation of the time. These measures included an agreement to put all mills on half time for sixty days in an effort to force prices up.[20] Similar paths were being beaten in the industry's other sectors. The tissue and manilla manufacturers met for the first time in Brattleboro, Vermont, and issued a call to its members to co-operate. The success of other associations figured prominently in its appeal:

The Writing Paper Manufacturers by frequent meetings and consultations relative to the demands of the market for their class of paper, have managed the supply and prices so as to yield a full percentage in their investments. Many other manufacturers, envelope, paper bag, tissue manilla, wire manufacturers, book publishers, etc. find it necessary to have similar meetings to consult and act together for their mutual interest. Therefore, believing that a better acquaintance and understanding among the manilla manufacturers would, at this time, promote their interests . . .[21]

The call went on to recommend the implementation of output restrictions.

Other evidence of a growing collective spirit among papermakers is found in the number of industry conventions that were beginning to take place, such as the one attended by thirty-three mills from five North

[18] *Ibid.*, p. 60; American Paper and Pulp Association, *The American Paper and Pulp Association: Constitution, Officers, Committees, Members, Historical Sketch* (New York, 1897) (hereafter APPA, *American Paper*), p. 43; Stevenson, *Background*, p. 18; McGaw, *Most Wonderful Machine*, pp. 194, 246.

[19] APPA, *American Paper*, p. 61. For more on this association, see Carlson, 'Associations', pp. 38–40.

[20] Smith, *History*, p. 196.   [21] *Ibid.*

Western states in Chicago in 1863, where samples of pulp made from straw, cornhusk, sorghum, and basswood were displayed.[22] Such meetings quickly became normal features of the trade in the United States. So much so that by the end of the century J. Luke, the president of the West Virginia Pulp and Paper Company, could remark that there had 'been conferences of papermakers as long as I have known anything about the business'.[23] These meetings were used as forums to raise debate about and facilitate co-operation on issues of common interest, as well as to publicise new technological breakthroughs and products. From one of these conventions in 1878 developed the American Paper Manufacturers Association, a nationwide body set up to represent the interests of the entire industry. With the initial intent of implementing co-operative measures to manipulate the market and control price movements, the association eventually took on a more social character, exercising its influence on work practices by aiding the dissemination of new ideas and by providing opportunities to its members for dialogue and intercourse on the problems confronting the industry. Innovation and expansion by members were also actively encouraged by the Association. Prizes for ideas useful to the trade were offered and the Paper Mill Mutual Insurance Company, an insurance company that guaranteed lower rates to papermakers, was established. The Association also took an interest in fostering the industry's exports, setting up a committee in 1883 to look into and oversee measures to that purpose. Its chief function, however, was not to circumvent competition, but rather to enhance it and make it more intelligent by removing unnecessary instability and disseminating facts. As the Association's president explained in 1914:

Our members are fully aware that this association is not for fixing prices or restricting production, but that it is entirely educational, with the sincere hope that prices will become more uniform through the knowledge of facts, and that the production will somewhere near equal the demand, due to the facts being developed by our statistics; also that our efforts have been consistently towards developing conditions, through efficient methods, that not only result in a benefit to the manufacturers, but to the consumer as well.[24]

Similar associations in individual branches of the industry were

22 Clark, *History of Manufacture*, vol. II, p. 36; Carlson, 'Association', p. 220.
23 *Pulp and Paper*, vol. III, p. 1483.
24 *PTJ*, 19 February 1914, p. 269. For more on the activities of the Association, see also *PTR*, 21 November 1883, p. 218; 11 January 1884, p. 333; 8 August 1884, p. 85; 21 November 1884, p. 325; and Carlson, 'Association', pp. 3–7, 281–2; APPA, *American Paper*, pp. 56, 60; *The Progress of Paper* (New York, 1947), p. 360; Weeks, *History*, p. 299. The association changed its name to the American Paper Manufacturers Association in 1883 and then to the American Paper and Pulp Association in 1891.

likewise established to protect the interest of their members and encourage the spread of best practice techniques. The Newsprint Manufacturers Association founded at the beginning of the twentieth century is one example.[25] These organisations played an important role in creating an awareness of common problems experienced in the industry and in spreading information on possible solutions to them. In 1862, for example, the mill-owner Addison Laflin delivered a lecture to the Writing Paper Manufacturers Association in which he attributed the industry's current problems to accounting practices that underestimated the true costs of production. In a display of openness that would have been unthinkable amongst British papermakers of the time, he supplied association members with detailed tables of the costs of production of various grades of paper made in his mill. He introduced the idea of process costing to those in attendance, differentiating for the first time between the costs of actually making the paper and the costs of its finishing.[26] The organisation proved to be a means by which the spread of accounting as well as technological innovation could be hastened. Such forms of collective invention, as Allen has called them, play a crucial role in the gradual accumulation of technology, especially in industries such as the paper industry, where the similarity of the technology used often led to the development of difficulties that were common to some extent to all firms.[27]

In this regard the American industry's experience was in marked contrast to the British whose association, the Paper Makers Association of Great Britain and Ireland, was politically and economically impotent and resembled more a fashionable club than a trade association. This inability of British papermakers to take concerted collective action stemmed not from the atomised nature of British production, as Lazonick has argued for the British industry in general,[28] for at this time American papermaking was equally fragmented and competitive, but from the prevailing business culture that regarded such associations as anathemas. John Evans, then president of the Association, admitted as much when he told the Royal Commission on the Depression of Trade and Industry in 1886 that he was unable to give the Association's opinions on the depression and its possible remedies because attendance at recent meetings, even committee meetings, had been poor. Moreover,

[25] Weeks, *History*, p. 300. For other similar associations, see Carlson, 'Association', pp. 62–4.
[26] J. A. McGaw, 'Accounting for innovation: technological change and business practice in the Berkshire county paper industry', *Technology and Culture* 26 (1985), 715–17.
[27] Allen, 'Collective invention'.
[28] B. Elbaum and W. Lazonick, (eds.), *The Decline of the British Economy* (Oxford, 1986), pp. 42–3.

he went on to note that many of those who actually did show interest were vehemently opposed in principle to the Association even discussing the possibility of intervention. This inability of the Association to consider the question of remedies to ease the difficulties faced by the trade perplexed Aird, one of the Commissioners:

MR. AIRD: Do you not think that, having warm interest in the welfare of the trade, that would be a responsibility which might be fairly met by the paper manufacturers? – that they should themselves as a body consider and determine that which is regarded as so desirable in the interest of the business?

JOHN EVANS: I think it would be very well if it could be arranged, that anything of the kind could take place; but we have in our body a vast variety of political opinion, and certainly in some cases there appears to be a view that any question relating to free trade is beyond the bounds of discussion, and that is a kind of deity which has been set up, and no blasphemy must be heard against it. That being the feeling with a certain number, of course there is a difficulty in getting any general consensus of opinion.[29]

It was a feeling that even prevented the Association from collecting statistics about the industry. On two occasions – in 1885 for the Royal Commission on the Depression and in 1904 for Chamberlain's Tariff Commission – statistics were requested, but on both occasions the Association failed to provide them. In 1904 less than a third of the industry even bothered to respond to the questionnaire the Association had sent around. By contrast, the American association had already successfully completed such an endeavour in 1883, and as was seen above, other bodies had done it even earlier than that.[30]

More important than the collection of statistics were the lost opportunities for the exchange of ideas that could have benefited the trade. Certainly there were papermakers in Britain who had grown tired of the 'languid after-dinner speculations of the Paper Makers' Club', as one critic described it, and felt that one of the main reasons why America was so successful was precisely because papermakers there bothered to pay attention and give material assistance to other members of the trade.[31] On the occasion of the twelfth AGM of the American

[29] Third Report of the Royal Commission on the Depression of Trade and Industry, PP XXIII(1886) (hereafter RC Depression), 587. For similar views expressed by George Chater, see p. 593.

[30] Ibid., p. 591; PMC, 10 November 1904, p. 411; 10 November 1883, p. 247; APPA, American Paper, p. 61. The American association set up a permanent statistics department in 1908. Carlson, 'Association', p. 288.

[31] PMC, 10 November 1883, p. 252; PTR, 2 November 1883, p. 181. Also see C. Arthur Pearson's after-dinner speech at the Paper Makers Association's Annual Dinner in 1904, in which he noted that, 'If a mutual exchange of ideas be acceptable and profitable to other people engaged in industrial pursuits, there can be no reason assigned debarring paper-makers. I would like to see their representative publications teeming with contributions as in others.' PMC, 10 March 1904, p. 92.

Paper Manufacturers Association in 1889, a contributor to the *Paper Trade Review* condemned in forthright terms the failure of the British trade association to emulate its American cousins, as well as its assumption that co-operative measures necessarily represented an abrogation of free trade:

> It is evident that American papermakers have RECONCILED themselves to meet upon a common understanding, without touching upon personal details or individual actions. The object is not to encroach in any way upon personal business management, but to promote a higher tone in competition and to encourage a firmer bond of unity in all matters of concern to the trade in order to facilitate its progress and permanence as a national industry.
>
> *Per contra*! How are the interests of the British paper trade being watched by the British Paper Makers' Association? Meetings are held periodically, but their conservatism is great. . . . There is no reason why paper manufacturers as a body should not have ideas in common in general tones of policy for the identification and protection of trade interests.[32]

Of course, this plea like others went unheeded, and when the First World War broke out in 1914, the British paper trade continued to be represented by an organisation that was not only ineffectual in speaking on its behalf in the political arena, but in even creating opportunities for the practical problems and suggestions of British papermakers to be heard by others in the trade. As such, a valuable conduit for the transmission of ideas and innovations through the industry was abandoned.[33]

Education

American associations, as well as individual producers, also saw the need for scientific instruction and research at a relatively early stage. By the turn of the century America had a system of technical and evening classes in papermaking that was more complete than, and in most cases, apparently far in advance of those in Britain. In many places the cost of these classes was defrayed or at least subsidised by the state government, while in others they ran on the bequests and gifts of wealthy philanthropists. Instances of the latter were the Arrow Institute of Chicago, the Pratt Institute of Brooklyn, and the Cooper Institute of New York. These establishments were all well equipped, and their

---

[32] *PTR*, 30 August 1889, p. 14.

[33] This American advantage continued into the twentieth century. An OEEC committee concerned with the post-war reconstruction of the industry throughout Europe lauded the American trade association for its continued contribution to 'the development of markets and the improvement of the American paper industry' and advised European producers of all nations to emulate their methods. OEEC, *The Pulp and Paper Industry in the U.S.A.* (Paris, 1951), p. 83.

graduates often immediately found employment in mills even if they had no prior work experience. Special arrangements were made for those in the more remote country districts, so that they could take courses by correspondence. By the beginning of the new century the International Correspondence School, for example, was already offering complete courses in papermaking by correspondence to American students.[34]

American papermakers also moved earlier to the establishment of institutions with an interest in research. In 1910 the first small research laboratory equipped with a grinder, barker, and wet machine was set up in Wausan, Wisconsin, primarily to examine and analyse problems associated with the manufacture and use of mechanically ground wood pulp. Two years later this was followed by a government forest products laboratory in the University of Wisconsin at Madison and the inauguration of the first collegiate course in papermaking at the University of Maine.[35]

Such institutions, however, supplemented rather than replaced those already existing at the firm level. Many American paper workers were fortunate to benefit not only from an ordinary day school education, but from technical instruction after work; a fact which made the American mill worker on average, according to Dyson, better educated than his or her British counterpart.[36] The Hammermill Paper Company of Erie, Pennsylvania, for example, took it upon itself to train and educate all of its employees, considering this the best way of ensuring a ready supply of skilled labour and consistently high quality workmanship.[37] In addition to such regular classes on technical and scientific matters other companies also found it beneficial to teach English to illiterates and new immigrants coming to work for them. Perhaps the most rigorous of these early undertakings to educate within the mill was that funded by the American Writing Paper Company in Holyoke. Operated very much like a vocational high school, this school held classes six nights a week and used a nearby abandoned mill as its laboratory. Though the numbers of those actually attending during our period is unknown, in 1917 this school had some 250 students.[38] Such schools, some of which had been running for a number of years, played an important role in raising the level of technical skill and understanding of the worker and presumably augmented the degree of innovation achieved through learning.

[34] *PMC*, 10 March 1903, pp. 96–7, 102–3; International Correspondence Schools Reference Library, *Sulphuric Acid, Alkalis and Hydrochloric Acid. Manufacture of Paper* (Scranton, 1902).
[35] Cohen, 'Economic determination', p. 3/11; Weeks, *History*, p. 301.
[36] Dyson, 'Paper Trade', p. 215.
[37] M. J. McQuillen and W. P. Garvey, *The Best Known Name in Paper: Hammermill. A History of the Company* (Erie, 1985), p. 29.
[38] Smith, *History*, p. 618.

The same type of technical education in papermaking was non-existent in Britain. As late as 1893, William Ross, the general secretary of the NUPMW, could claim that he was unaware of any such classes operating in the country and that to his reckoning at least 80 per cent of mill workers had no training at all for their jobs.[39]

The problem started with the young. Lord's report on the paper industry for the Fourth Report of the Commissioners on the Employment of Children described the low educational standards of children employed in paper-mills, some of whom did not know whether England was governed by a king or a queen, had never heard of Queen Victoria, and could not recognise the names of major towns in their own county.[40] Some papermakers, such as Alexander Cowan and Son and Hele Paper Works, did provide rudimentary educational facilities and insisted upon its workers having a modicum of education, though these mills never numbered more than a mere handful. The motivation for these classes was not purely paternalistic, for, as Collins of Hele Paper Works argued, it was also an investment that would pay off when the youngsters matured and were able to be entrusted with more responsibilities. It needs to be remembered, however, that the emphasis of these classes was always on reading, writing, and arithmetic (and sewing for the girls), and not technical instruction, so that the only papermaking a new employee could learn in even these mills would be whatever he could pick up from his elders.[41] This was often not the most efficient channel of transmitting information across the generations, especially when the older workers were not interested in helping the younger workers out or themselves understood very little about the work they were doing. As one young papermaker complained to the *Paper Trade Review* in 1884, relying on fellow workers for information was pointless: 'Now, as far as my experience goes, most of the men either return no answer or give one that has not a grain truth in it, if you ask them about the stuff they are working; but, if you ask them which horses ran the Derby, they give a truthful answer.'[42]

There was thus little opportunity for the young, conscientious paper-mill operative to improve himself through education. In the same year another sixteen year old suggested to the same journal that paper-mill owners ought to organise classes and essay competitions on paper-making for the boys in their establishment. This would not only be

---

[39] *RC Labour*, pp. 510, 749. Dyson could claim the same thing ten years later, 'Paper trade', p. 215.
[40] *Fourth Report on Employment*, p. 299.
[41] *Ibid.*, pp. 328–9; Cowans Collection, GD 311/1/9.
[42] *PTR*, 13 June 1884, p. 683.

beneficial to the boys, but 'would pay mill owners in the end'.[43] That these suggestions should occur at this time is not surprising. This was, after all, the time of the Royal Commission on Technical Instruction, and talk of reforming the existing methods was openly mooted in even the paper trade.[44] But from those who believed that there was no substitute for experience, the resistance to change was strong:

Take a machine man, for instance. He may be a man who understands all about the machine, in theory; but put him on a machine going a good speed, and should the paper break for an hour or so, he will probably get so excited that he can scarcely tell the wet from the dry end of his machine. Such men are frequently to be met with, and I dare say, would be able to pass a rigid examination where many practical men would not.[45]

From 1884 the City and Guilds of London Institute began offering courses in papermaking, though these struggled from the beginning because of lack of interest. In the 1899/1900 academic year there were only two registered papermaking classes with thirty-two students in the country. Interestingly, most of these students were not practical paper-makers.[46] As the realisation that, just perhaps, the British paper industry was lagging behind the American, began to sink in, the Paper Makers Association set up a committee headed by Lewis Evans and Richard Barton to look afresh into the question of technical education. By October 1899 this committee had published its report, suggesting that a revamping of the existing organisation and structures with appropriate incentives, rather than the intervention of the Association, was the appropriate course of action. It attributed the failure of the City and Guilds' course to the fact that it was too theoretical in content and because it could not cater for the needs of all the papermaking regions of the country. To rectify these defects the committee recommended that courses and examinations be written so as to have a distinct practical bent, that the association offer a prize of £1 to those paper-mill workers who got an honours grade and 10s for a pass, and that paper-mill owners be urged to take it upon themselves to organise technical classes in their area and encourage their workers to enrol in them.[47]

The committee had some marginal, if fleeting, successes. New classes

---

[43] *PTR*, 6 June 1884, p. 667.
[44] For example, see *PTR*, 1 February 1884, p. 379; *PMC*, 10 March 1887, p. 76. The Royal Commission itself did not report on the paper industry, but the letters of two paper hanging producers were received, both of whom could see no value in technical instruction except that it might improve the tastes of the masses. Second Report of the Royal Commission on Technical Instruction, *PP* XXXI (1884), 668.
[45] The views of one papermaker reported in *PTR*, 15 February 1884, p. 441.
[46] *PTR*, 28 November 1884, p. 337; *PMC*, 10 November 1899, p. 513.
[47] *PMC*, 10 October 1899, p. 436.

were inaugurated at the Manchester Technical School, the Science and Art School in Darwen, the Municipal Technical School in Bury, the Technical School in Maidstone, as well as in Dickinson's mills in Hemel Hempstead and Watford. All in all, 231 students were enrolled in papermaking classes in 1901.[48] The most important and lasting of these schools was at the Manchester Technical School where even a small twenty-four inch Fourdrinier constructed by Messrs. Hemmer and Brotler of Weidenfels was installed for use by the pupils.[49] This, however, should not be mistaken for a school intended for the average worker in the mill. To enrol students had to have already attained a fairly high standard of education and experience. Moreover, as J. Reynolds, the school's principal, explained, the 'students they wanted in the technical school were not the paper-mill operative, but the sons of owners and managers – the future captains of industry'.[50]

As it turned out, most of these new classes were shortlived. Attempts to spread them to the south of the country were also met by a wall of apathy. In 1902 the principal of Battersea Polytechnic sent out letters to 75 mills and 85 wholesale stationers and paper-mill agents in the London area enquiring whether they would be interested in the technical classes in papermaking that the Polytechnic was thinking of starting up. Out of these 160 letters only 6 replies were received, three of which were in favour of the scheme.[51] By the following year the Association's technical committee, faced by the faltering numbers of classes and students in the country, acknowledged the failure of its policy, and from that date little more was heard from the committee.[52] As a result, technical instruction was not to make any further headway in Britain in the years leading up to the First World War.[53]

The core of the problem was that papermakers remained unconvinced of the benefits of technical classes and in many instances were even unaware of their existence. The prevailing view was that all that workers needed to know could be acquired on the job. The comments of

---

[48] *PMC*, 10 May 1901, p. 201; 10 May 1902, p. 170.
[49] British paper-machine makers had refused to make the machine as they considered it a 'toy'. *PMC*, 10 July 1902, p. 246.
[50] *PMC*, 10 October 1902, p. 377. Also see the similar comments by one of the lecturers, Julius Hübner, in *PMC*, 10 February 1902, p. 46.
[51] *PMC*, 10 October 1902, p. 378.   [52] *PMC*, 10 March 1903, p. 93.
[53] Its progress after the war was not amazing either. In 1942 Meredith could still note, 'there is, in fact, no known way of training paper makers except by making paper: on the one hand it is not an apprenticeship craft, nor are there technical classes for learning it; on the other hand, skill in it cannot be picked up rapidly, it may take six or seven years to acquire anything like the mastery: the inexperienced man may be a danger both to himself and to his mates, and inflict great loss upon his employer'. Meredith, 'Impact', pp. 9–10 in Social Reconstruction Survey, C14/119.

A. Poulter of London Paper Mill – one of the biggest papermakers in the country – are indicative of the widespread belief of a great many in the trade at the time, and in scepticism echo the sentiments made twenty years prior by the papermaker quoted earlier: 'I think that the bulk of the workmen are as well educated as is necessary for the actual making of paper, which is a mechanical process. I do not think a man requires a college education to fill a beater, or even to run a machine.'[54]

Poulter, like so many of his peers in the trade at this time, failed to realise that technical education might not only teach workers how to perform specific tasks, but also to understand and resolve problems that occurred in the day-to-day running of the mill. As a consequence technical education in British papermaking remained undeveloped, and a means of enhancing performance and technological accumulation was forfeited. This was no better demonstrated than by the difficulty even very progressive firms such as Edward Lloyd Limited had in running their new machines at optimum speed. As Neil Turner, one of the supervisors, explained: 'the difficulty is not with the machines themselves, but with the men. The men want a certain amount of training to tackle machines running at enormous speeds.' The firm then engaged in a gradual process of bringing their workforce up to scratch.[55] If we add these types of inefficiencies to the potentially useful technological knowledge that was lost as a result of ignorance, the lack of proper training within British mills probably cost the industry dearly.

Firm size and structure

Attempts to control the market in America extended beyond official organisations. The merging of firms and the production of ever-larger mills was a major feature of the American industry from the end of the century onwards.[56] Although between 1890 and 1913 this trend was still in its early stages, it had already progressed far enough by then for it to start having some influence on the rate of technological accumulation achieved in the United States. This was particularly so amongst the largest of the American firms which were able to attain the long runs and continuous production that were most susceptible to the economies of practice.

[54] Tariff Collection, TC3 1/89, p. 7. Similar views were expressed by Evans, TC7 28/2 E(4); Dixon, TC3 1/86 p. 10; and Green, *PMC*, 10 March 1903 p. 99.

[55] *PMC*, 10 December 1900, p. 551.

[56] As a result Dixon was induced to remark of American papermakers that, 'they think nothing of starting to build a mill and putting six to eight machines in it. To an Englishman, it is staggering to see eight machines in one house; when we think of our little houses with two machines in them.' Tariff Collection, TC3 1/86, p. 5.

As Weeks noted in 1916, whilst reflecting on the moves towards even greater combinations that were going on around him, 'disposition had long existed towards loose gentlemen's agreements so-called, in nearly all branches of the industry'.[57] This movement towards combination, however, only started to reach fever pitch between 1890 and 1914.[58] Reasons for the movement lay in the significant economies of scale available to those who could expand production, the vast amounts of capital investment needed to modernise and attain this scale, and the ever-present desire of manufacturers to maintain prices and profits through the exercise of market power.[59] Beneath all this, however, lay the rapidly growing demand for paper in America, without which it would not have even been possible for paper manufacturers to plan such bold expansions.

As a result of these expansions the industry advanced as it had never before. Many concerns were incorporated with capitalisations of as high as several million dollars, while other pre-existing ones increased their capitalisation by similar amounts. Average firm size measured in volume of output increased by over 200 per cent between 1895 and 1905, while average mill size also grew by 121 per cent over the same period.[60] Also indicative of the growth in the size of concerns was the capital investment needed to set up an average size mill which leapt enormously from $16,390 in 1849 to around $744,603 in 1914.[61] This growth in average size was most evident in the newsprint branch of the trade, which between 1895 and 1905 grew at a staggering 224 per cent, or at an annual compound rate of 14.1 per cent, from 45,400 pounds per day to 169,800.[62]

This was also a time of greater concentration within the industry. By 1895 only twenty-two firms, or 3 per cent of the industry, were responsible for a quarter of the industry's total output. Between 1897 and 1907 as much as 22 per cent of the industry's firms disappeared through mergers, though this figure was even higher in the newsprint (47.4 per cent), and book and writing paper (36.9 per cent) sectors. Another characteristic of the period was the domination of large combinations, in that a few consolidations, each already composed of several companies of above average size, were responsible for a comparatively weighty share of the total number of acquisitions and of

[57] Weeks, *History*, p. 302.
[58] Cummins, 'Concentration', pp. 36, 130, 136; Lamoreaux, 'Industrial concentration', ch. 4.
[59] Lamoreaux, 'Industrial concentration', pp. 393–4.
[60] Cummins, 'Concentration', pp. 38, 42; Weeks, *History*, p. 311.
[61] Cohen, 'Economic determination', p. 3/11.
[62] Cummins, 'Concentration', p. 45.

the total capacity procured. Moreover, the vast majority of the mergers (93.4 per cent) were of firms producing the same type of paper.[63]

The conspicuous success of the combination movement was the International Paper Company (IPC). Incorporated in 1898 it acquired sixteen mills manufacturing newsprint in five eastern states and gradually added other paper- and pulp-mills, woodlands, and water-power to its possessions. With thirty-four mills under its control in 1900 its assets amounted to $41,586,964 in mill plant and $4,101,723 in woodlands and other property.[64] The company was not only large, but broke all family ties by hiring professional executives, and moved to centralise its administration, rationalise its production, and develop long-term strategies of vertical integration, product diversification, and demand creation.

In this period virtually all papermaking firms like the International, that wished to produce their own pulp, opted for the establishment of new pulp-mills and the acquisition of their own timber sources rather than merging with existing concerns.[65] In 1898 the IPC owned half a million acres of spruce woodland in New York, Michigan, and various New England states, and through its policy of active acquisition of further supplies this figure had more than doubled by the end of the next year. Likewise, around this time, the St Regis Paper Company acquired water rights and 98,000 acres of woodland in upstate New York.[66] These acquisitions circumvented a problem that plagued many British papermakers who relied on others for their supplies of raw material. The Scottish firm, Messrs Brown and Co., for example, had an on-going problem with its supplier which interfered with the regular flow of material through its mill. In response to the disruptions caused by the irregular arrival of its consignments of esparto, the firm complained to its supplier:

We are very averse to having this extra quantity, and when we telegraphed about the delivery by the voyage, mentioned 225 tons as the extent we wished, and it is the contract quantity. In the circumstances you mention and upon this occasion we do not care to return acceptance, but we shall have more Grass than we require and being also short of room it will be altogether inconvenient for us . . . In absence of your deliveries some time ago we had to buy elsewhere and we have some also of these lots in stock.[67]

The IPC, like a growing number of other American firms, was also

[63] Ibid., pp. 80–98.   [64] Weeks, History, p. 302.
[65] Cummins, 'Concentration', p. 99.
[66] Amigo, Nueffer, and Maunder, Beyond the Adirondacks, p. 43.
[67] Brown Collection, GD1/575/8, p. 201. It was also a matter of the quality of the material received. See Cowans Collection, GD 311/2/45, p. 272.

active in creating its own sales department with agencies in all the principal cities both at home and abroad. These were charged with the responsibility of boosting sales. In January 1899, for example, the International established an office in London headed up by Smart, formerly a manager at Messrs Thomas Owen and Company Ltd. of Cardiff.[68] British producers, however, generally chose not to concern themselves with the distribution of their output, leaving this almost exclusively to independent agents and paper merchants.[69] This strategy was not without its advantages, as it shifted a lot of the risk and burden of distributing one's own products at a relatively modest cost – between 2.5 and 5 per cent of the profits. Moreover, given the expertise and detailed knowledge of the different markets for paper that paper merchants possessed, this mechanism could work very efficiently indeed. But as it became increasingly necessary to dispose of larger and larger volumes of paper, or alternatively when the firm hoped to branch into another sector of the trade, the existing system with all of its transaction costs became less acceptable to the papermaker. As William Todd of Brown and Co. told a paper trader in London, who wanted the firm's business, in 1877, the prosperity of the mill depended heavily on the efforts of its agent whose actions were largely unknown and unaccountable to the firm:

We would like very much if you could find out for us the wholesale houses and others for whom Mr McMurray [the mill's sole agent] calls or to whom our papers are sold by him regularly. We cannot just ascertain this from himself or his people and want very much to know if we are well represented to the trade and if we see any way of helping you and ourselves with houses out of Mr. McMurray's beat, we may think of trying to get in with them in this outside way.[70]

The IPC's policy of procuring woodlands and investing part of its profits in longer-term development projects enabled the company to acquire and retain for itself, despite intense political and economic opposition, a prominent position in the industry. By 1901 the IPC's share of the newsprint market had reached 65 per cent, although this fell somewhat in subsequent decades, especially as a result of later competition from Canada. The firm evinced in particular a strong commitment

---

[68] *PMC*, 10 January 1899, p. 23. He was replaced by William Sinclair formerly of Edward Lloyd Ltd. later that year. *PMC*, 10 August 1899, p. 349. For other firms' sales agencies, see Carlson, 'Association', pp. 141–2.

[69] W. Reader, *Bowaters: A History* (Cambridge, 1981), pp. 9–10. For examples, see Ketelbey, *Tullis Russell*, pp. 6–7; Weatherill, *One Hundred Years*, p. 5; and CKS: Hollingworth Collection, U1999 B8/8.

[70] Brown Collection, GD1/575/8, p. 13. In 1880 the firm finally changed its agents and began accepting the occasional direct order. GD1/575/8, p. 121.

to technological change and regularly put its money into experimenting with new technologies and strategies for future growth. It also operated a mill at Glenns Falls and a tree nursery in Vermont where experiments with new fibres, grinding machinery, and forestry techniques were tried out before being used in the firm's production mills. In 1901 alone the firm spent a million dollars on such improvements in plant and equipment.[71]

Companies like the IPC formed the vanguard of a movement that was to reorganise and modernise the industry in America as the new century unfolded.[72] Before the First World War, however, the movement still remained very much in its infancy with only one in five papermaking firms producing some of their own pulp at the end of the first decade of the twentieth century.[73] Moreover, average mill size in the States remained roughly the same as in Britain right up to the turn of the century, at which time the impact of the combination and merger movement began to take effect. Still, by 1914, the gap between the two nations in terms of the scale of production attained in the average mill did not exceed the 10 per cent mark. As such, one cannot attribute too much of America's labour productivity lead in the Edwardian period to the greater throughput that integration permitted.

Despite this, these large and modern American firms did play a very important role in bringing about a new degree of awareness in the trade of the need for innovation to remain viable. More importantly, given their key position in the industry, the longer, more specialised runs that these large firms possessed provided them with abundant opportunities to benefit from the economies of practice and to play a disproportionately large role in creating the massive increases in machine speeds of the decades leading up to the First World War. Indeed, many of the technological breakthroughs needed to bring about this surge originated

[71] Vandermeulen, 'Technological change', p. 265; Smith, *History*, pp. 201–3. Many companies also had a policy of purchasing all valuable inventions in their branch of the trade, so that it could develop them for their sole use. This kept out competition and ensured that good ideas were always quickly taken up. Out of the some 300 patents acquired by the Union Bag and Paper Company of Chicago since 1871, over half were still being actively used in the company's mills in 1896. *PMC*, 10 January 1896, p. 11; Carlson, 'Association', p. 64.

[72] The IPC was not the only big combination: it was only the best known. Others combines included *inter alia* the Great Northern Company (1899), The American Writing Paper Company (1899), United States Envelope Company (1898), Union Bag and Paper Company, United States Paper Bag Manufacturers Association, American Straw Board Company (1889), and the National Straw Board Company. Many of these were unsuccessful and had to disband voluntarily at a later date. For more on these and the IPC, see Chandler, *Scale and Scope*, pp. 112–13; Weeks, *History*, pp. 302–13, and Keir, *Manufacturing*, pp. 492–4.

[73] Ohanian, *American Pulp*, p. 22.

or were developed in the mills of these larger establishments. Furthermore, by setting the technological pace, especially in the newsprint
sector, where they predominated, these larger, integrated firms forced
the rest of the industry to follow suit to stay competitive.
Although in 1913 the American industry still had a long way to go to
make vertical integration a norm of the American paper trade, it was
nonetheless a path that the Americans had gone further down than their
counterparts in Britain. The British paper industry of the time had little
to compare with America's large, integrating, professionally run concerns. The Wall Paper Manufacturer Company formed in 1899
managed to acquire a large share of the British wallpaper market,
whereas the entrepreneur Edward Lloyd, publisher of the *Weekly News
in London*, not only built his own paper-mill in Sittingbourne, Kent, but
also acquired 180,000 acres of land in Algeria and timber-mills in
Norway, where he produced the esparto grass and wood used in the
manufacture of his paper, and which he brought over to England on
board his own ships. Such cases, however, were certainly not the norm.
At any rate in terms of scale these British efforts paled in comparison to
the large American establishments. In 1910, for example, Edward Lloyd
Ltd., even though it was the biggest paper manufacturer in Europe,
produced only one-seventh of the IPC's annual output.[74]

Labour relations

Given that a significant proportion of technological change in the
industry originated in the ideas and innovations of those who worked in
it, industrial relations can play an important part in determining the
degree of technological change that is realised. Working in an atmo-
sphere of industrial conflict, characterised by friction between labour
and capital and resentment of the owner, a worker may simply have no
desire to help the boss out by improving his machinery; at least, not
unless something is given in return. This was not lost on many American
papermakers. In fact, American mill owners on the whole seemed to
have appreciated better than British the necessity of providing a suitable
work environment and conditions for their workers, and of instilling

[74] J. Munsell, *Chronology of the Origin and Progress of Paper and Paper-making* (New York,
1980 [1876]), p. 221; and *PMBPTJ*, November 1902, p. 15. In 1910 Lloyds produced
67,500 and the IPC 470,000 tons. See, *Pulp and Paper*, vol. V, p. 3294. Dickinsons also
was a very progressive firm with export managers and representatives overseas, though
it must be remembered that the firm had been paper stationers and had already
employed some of these devices even before it had begun to make paper for itself. See
Evans, *Endless Web*, *passim*, and L. Evans, *The Firm of John Dickinson and Company
Limited* (London, 1896), p. 21.

them with positive and desirable attitudes to technological change. As an American papermaker pointed out in 1887, care, knowledge, loyalty, and intelligence – and not long hours and low wages – were the essential ingredients of success in papermaking.[75]

Another important ingredient was the nurturing of a sense of justice, equity, and equality in employees. To a large extent such a feeling seems to have existed and been shared in the American industry by workers and owners alike. In part this feeling stemmed from shared experiences and activities, like the town baseball team and fire company, where worker and employer played and worked side by side; but in part it was also related to the genuine belief held by many that one could progress through the ranks with hard work and diligence from assistant to shareholder. Even in the first decade of this century, there was still evidence to support the retention of such a belief. David Clark Everest, for example, advanced from office boy in 1900 to assistant manager and then partner of Marathon Paper Mills Company in Wisconsin in just nine years. Lou Calder did the same at Perkins-Goodwin. Likewise, on his tour of the United States, Dyson not only came across many American workers who shared this view, but also found British paper-makers who had left their homeland because of the better opportunities and conditions on the other side of the Atlantic.[76]

The paternalism of American mill owners, however, went beyond good pay. Many of them also took an active interest in the health of their workers by ensuring that they were vaccinated and that mills were always kept both salubrious and comfortable.[77] American mills were thus generally better lit, ventilated, and safer than British mills. Ventilation was particularly impressive in America. As Dyson told the American *Paper Trade Journal* in December 1902:

I'll venture to say there is not a single mill [in Britain], Lloyd's inclusive, in which ventilation is given much attention. Of course, there are exhaust fans to remove the steam. They also remove the heat but the men work in discomfort and are therefore not to be blamed for not doing so much work as do your men. They are working under disadvantages and foul air tires a man very quickly.[78]

But such attitudes and practices in America need not be attributed purely to the munificence of the papermaker there. A decent standard of living, job security, and the opportunity to have grievances heard in open

---

[75] Green, *Holyoke*, p. 210. For other examples, see *Helps to Profitable Paper-making*, pp. 78–83; and McQuillen and Garvey, *Hammermill*, p. 10.
[76] Boese, *Papermakers*, pp. 66; Karges, 'David Clark Everest', p. 28; Calder, *The First Hundred Years*, p. 42; McGaw, *Most Wonderful Machine*, pp. 319–28; *PMC*, 10 January 1903, p. 26.
[77] *PMC*, 10 April 1884, p. 94.
[78] *PMC*, 10 January 1903, p. 25; also 11 June 1894, p. 278.

discussion were regarded by many employers as essential to social cohesion as well as to business prosperity in the long run. These were also effective means of keeping unions out the mill.[79] In a perverse way the greater threat of union activity in the United States seemed to have made its papermakers more keen to explore other means of dealing with the demands of its workers than confrontation. It certainly had its effect and led, according to one commentator, to a rate of labour turnover in the paper industry that was exceptionally low.[80]

American firms were also very active in experimenting with various fringe benefits for the workers such as free technical instruction and the establishment of mutual relief and rudimentary insurance associations. More importantly, a system of giving premiums, bonuses, and promotions for improvements and suggestions made by an employee was largely used in the United States and frequently brought beneficial results all round. In the earliest issues of the *Paper Trade Journal*, the American industry's most influential publication, readers raised the question of bonuses as an effective means of inducing the interest of employees in their work. Over the coming years interest in such schemes continued to spread rapidly, such that by the 1890s many firms in the eastern states had profit-sharing plans already in place.[81]

The firm S. D. Warren Co., based in Congin Falls, Maine, was the exemplar of these firms.[82] Starting with 50 employees in 1854 the company expanded and integrated, acquiring a wood pulp-mill in Yarmouth by 1875, and by 1888 had 990 workers on its books. An important aspect of its success was its enlightened treatment of its employees. Portland's *Eastern Argus* wrote of the firm that 'here friction between capital and labour is unknown, affording the practical example of the true solution to the labour question'.[83] The firm was also very active in the community, providing a building site and $5,000 towards the construction of the local church, as well as contributing funds for halls, libraries, and public reading rooms in nearby Westbrook. It was also genuinely concerned with the welfare of its workforce. Money was loaned to employees for the purchase of homes, payment of bills, or the education of children, and the terms of repayment the firm offered were liberal, allowing the loan to be paid off interest free by minor deductions from the weekly pay packet. In the early 1890s it was also one of the first mills to introduce the three-tour, eight-hour day. The policy of the firm

---

[79] Karges, 'David Clark Everest', pp. 195, 218.
[80] Green, *Holyoke*, p. 100. See also *PMC*, 10 February 1899, p. 55. Voelker, 'History', p. 75 also says that although the occasional strike did take place, cordial relations were the norm.
[81] *PTJ*, 2 December 1872; Smith, *History*, pp. 594–5; Dyson, 'Paper trade', *passim*.
[82] Smith, *History*, pp. 159–64.   [83] *Ibid.*, p. 162.

at all times was to encourage workers to take an active interest in the company. A central part of the firm's attempt to accomplish this was the profit participation plan it implemented in 1891. This plan provided that those employees who worked a minimum of 75 per cent of the year's working days would receive a dividend determined by dividing a percentage of the net earnings in ratio to the total earnings of the employees. The company made it clear that the intention of this plan was to enhance the competitiveness and well-being of all in the firm and urge the men to use economy in their work and not fear new machinery or new practices. As S. D. Warren explained to J. E. Warren in a letter dated 12 March 1891:

Under the plans now proposed, it will be for the interest of these men to work as he would on his own account; and a just regard for his own interest would make it right for him to point out to the management any failure in duty in efficiency on the part of others.[84]

The plan proved to be a success, as a result raising productivity and the average wage in the mill immediately by between 2.5 and 7.5 per cent.

The actions of S. D. Warren Co. were watched and emulated by other mills in the country, so that by the end of the century the practice of giving bonuses and even sharing profits were quite widespread throughout America.[85] Although there were undoubtedly exceptions to this generalisation, Dyson could still conclude in 1902 that as a general rule in the United States every encouragement was given to the worker 'in the shape of reward for initiating or suggesting any improvement in the management of machinery and in the event of the production being increased thereby, the workmen may rely upon getting a fair percentage of the profit therefrom, whether he is on piecework or day wage'.[86] The result was a closer relationship between employer and employee, and technological change came to be 'appreciated by the workmen for

[84] *Ibid.*, p. 164. S. D. Warren Company, *A History of S. D. Warren Company, 1854–1954* (Westbrook, 1955), *passim*.
[85] Among the other firms who employed bonuses for increased production and efficiency and expressed satisfaction with them were the United Paperboard Co., Dennison Manufacturing Co., Detroit Sulphite Pulp and Paper Co., Fox Paper Co., Peninsular Paper Co., Crocker and Burbank Co., and National Paper Co. National Civic Federation, *Profit-sharing by American Employers* (New York, 1920), pp. 60, 102, 148, 180; National Industrial Conference Board, *Practical Experience with Profit-sharing in Industrial Establishments* (Boston, 1920), pp. 83–6.
[86] Dyson, 'Paper trade', p. 215. Although it is not possible to ascertain accurately the total number of American cases, because of the lack of comprehensive surveys there, Dyson was probably right in his assessment of the greater prevalence of such practices in the United States. N. R. Gilman, *Profit-sharing between Employer and Employee: A Study in the Evolution of the Wages System* (New York, 1889), pp. 390–1 and M. W. Calkin, *Sharing the Profits* (Boston, 1888), p. 14, for example, also both found more instances of profit-sharing in the United States than in Britain in the 1880s.

reasons that up-to-date machinery means less anxiety and better results for both'.[87] One would expect this to be a much healthier environment for innovation than in many British mills where suspicion between capital and labour, as well as of change in general, was more usually the case.[88] Certainly Dyson believed so. Asked what aspect of American practice should be introduced in Britain, he replied, 'improved machinery to run at high speed, rewards for the encouragement of suggestions and efficient ventilation and sanitary arrangements'.[89]

Dyson's report attracted a lot of attention within the trade and was covered by all the major trade journals.[90] In February 1903 he was asked to deliver the keynote address at the Annual Dinner of the Paper Makers Association. Following this address a lively discussion and frank exchange of ideas ensued. One consequence of that discussion was that the Association's Annual Report of the year included an appeal to those papermakers who were prospering to reveal the secrets of their internal administration.[91] By the time of the AGM in June of that year some progress in that direction already seems to have been made. At this meeting the creation of an Employment Suggestion Register, whereby the Association could reward useful suggestions that workmen had passed on to it, was recommended. It was also suggested that the same scheme be tried by individual mills as 'it had borne splendid fruit in America'. The members were then told of the endeavours along these lines that had already been made in Lloyds' mill in Sittingbourne. At this mill a placard was posted for public display which read:

If employés engaged in any branch of work in any of these mills can see a better, quicker or more economical way of doing it, or can think of some improvements that might be made in the work; or if any idea should occur to them that could be of use in the business, they should write their suggestion in ink, giving their full name, occupation and address, with date, and drop it into the suggestion box at the entrance lodge which is kept under padlock and key and which will be opened at stated periods by the manager, when their suggestions will be carefully considered and if, found of practical utility, prizes will be awarded according to merit. One person may make as many suggestions as he pleases.[92]

A paper worker attending the AGM, however, voiced grave doubts

[87] Dyson, 'Paper trade', p. 218.
[88] Typical of that mistrust was the dispute that took place at Messrs Smith, Shore, and Knight in Birmingham in 1886. The firm reduced wages by a third, and when angry workers threatened to strike, they were promptly replaced by unemployed workers from Birmingham and elsewhere. *PMC*, 10 April 1886, p. 120. For similar disputes see *RC Labour*, p. 759; *PTR*, 5 December 1884, pp. 385–9.
[89] Dyson, 'Paper trade', p. 218.   [90] For example, see *PMBPTJ*, November 1902.
[91] *PMC*, 10 March 1903, pp. 93–101.   [92] *PMC*, 10 June 1903, p. 203.

about the efficacy of such a scheme. The problem, according to this worker, was that a belief prevailed amongst British workmen that employers could not be trusted to treat them fairly in such matters. Moreover, employers in Britain seemed to regard workmen as theirs – 'body and soul' – and that any useful idea a worker might have was their sole property. He illustrated these views by relating an experience of his own from his first trip to America. On that trip the English firm, for which he was working and to which the visit yielded a handsome profit of £1,000, declined to insure his life for the journey or reward him for his exceptional work; on the other hand, an American firm, to which he gave some assistance whilst there in getting a machine running, treated him very kindly, insisting that he accept a bonus of £100 for his troubles. It was the first time he had ever handled such a sum of money, and it led him to claim that 'if employers in the paper trade would deal fairly and liberally with their workpeople who made valuable suggestions, in a few years this scheme would prove highly beneficial to the trade'.[93] However, as with so many other projects initiated by the Paper Makers Association, the Employment Suggestion Register was quickly dropped because of the number of complaints received from members who disapproved of their membership dues being wasted on expenditure that was deemed both unnecessary, and potentially an interference of the private management of individual businesses. It was thus left up to each individual mill to find the best way, if any, of stimulating their employee's interest in boosting productivity. Most of them at this time chose not to do anything.[94]

Of course, the question of incentives for labour was not new to British papermakers in 1903. By then already there had been a long history of profit-sharing, both in Britain and elsewhere. Lord Wallscourt, for example, is reputed to have experimented with various schemes on his 100 acre farm in England sometime between 1829 and 1832. Most of these early instances of profit-sharing, however, appear to have occurred in either France or the United States. In any case, for the time being, such efforts failed to generate any great deal of intellectual or commercial interest in the practice; for although Charles Babbage did discuss it at some length, the first detailed historical and theoretical investigation of profit-sharing was in fact only carried out by the German

[93] *Ibid.* p. 204.
[94] *PMC*, 10 March 1904, p. 84. For example, leading firms William Sommerville and Co. Ltd. and Thames Board Mill Ltd. only introduced bonus schemes after the Second World War. These proved successful. Watson, *Last Mill*, p. 66 and British Productivity Council, *The British Productivity Council Case Studies 2: Plant Maintenance* (London, 1956), pp. 42–5.

academic Victor Böhmert in 1878.[95] In its wake Böhmert's work
initiated an unprecedented investigation of the subject, so that by 1889
when the first International Congress on Profit-Sharing took place in
Paris, its advocacy had already burgeoned into a movement of international proportions. Soon thereafter treatises detailing numerous case
studies were published by *inter alia* Calkins, Gilman, and Rawson.[96] To
exponents profit-sharing represented the only 'ethical and economical
solution to the labour question', and, in turn, promised to grant
employers the co-operation and loyalty of their workforce, a reduction in
rate of labour turnover, economies of material, greater productive
efficiency, fewer industrial disputes, and the promotion of thrift. All of
this was possible because through profit-sharing the employer could
'offer a keen incentive to the ambition, fidelity, and industry of the
labourer, and thus actually [create] an entirely new source of profit'.[97]
The intellectual climate generated by these publications even induced
the Board of Trade to take an interest in the idea, and between 1890 and
1914 it compiled five reports on profit-sharing and labour co-partnership
in Britain and abroad.[98] The paper trade also was certainly not immune
to the advocates of profit-sharing. As early as 1889 an article had
appeared in the *Paper Maker's Circular* on the practice of profit-sharing
in European mills. In this article attention was given to companies, such
as the French cigarette papermakers Messrs Abadie and Co., which
shared up to an eighth of its annual net profits with its employees. Since
the implementation of their scheme nearly a decade earlier, the firm had
boosted its productivity and was clearly very pleased with the result of its
policy:

Thanks to the powerful stimulant the whole workforce has never ceased to be
rivals in zeal in the work in order to obtain the largest possible result: cases of
individuals indifferent to all the welfare which one could procure for themselves
have been extremely rare and we may say to-day that the whole force has been
for long years in service of their house.[99]

[95] T. J. Hatton, 'Profit sharing in British industry, 1865–1913', *International Journal of Industrial Organization* 6 (1988), 69–90. C. C. Balderston, *Profit Sharing for Wage Earners* (New York, 1937), pp. 7–9; Dr V. Böhmert, *Die Gewinnbeteiligung. Untersuchungen über Arbeitslohn und Unternehmergewinn* (Leipzig, 1878).
[96] Calkins, *Sharing*; Gilman, *Profit-sharing*; H. G. Rawson, *Profit-sharing Precedents* (London, 1891). Other early studies include NCF, *Profit-sharing*; NICB, *Practical Experience*; and J. A. Bowie, *Sharing Profits with Employees* (London, 1922).
[97] Calkins, *Sharing*, pp. 9–12.
[98] Board of Trade (Dept. of Labour), *Report by J. Lowry Whittle on Profit-sharing* (London, 1890); *Report by D. F. Schloss on Profit-sharing* (London, 1894); *Report on 'Gain-sharing' and Certain other Systems of Bonus on Production* (London, 1895); *Report on Profit-sharing and Labour Co-partnership in the United Kingdom* (London, 1912); and *Report on Profit-sharing and Labour Co-partnership Abroad* (London, 1914).
[99] *PMC*, 10 October 1889, p. 374.

In the British paper industry such systems of co-operative production bonuses and profit-sharing were very rare.[100] This contention can be verified, for, unlike in America, the Board of Trade did collect fairly reliable statistics on profit-sharing in Britain from 1890. In that year the only instance of the practice ever recorded in the paper trade was at Hele Paper Works, Devonshire, which had introduced a bonus scheme in July 1889. By 1912 still only five attempts in all had been tried in the British industry, of which just four remained operative. Together these surviving efforts involved some 794 employees, or a mere 1.75 per cent of the workforce.[101] More common than profit-sharing were *ex gratia* payments made to senior members on retirement and on holidays such as Christmas. For instance, Hollingworths provided Christmas bonuses and presents to most of its employees every year. One's behaviour and circumstances usually determined the generosity of the gifts. Union members generally received less, whilst those facing particular hardships fared better. For Christmas 1885, for example, Mrs Clegg, a raghouse worker, not only got the standard 2s 6d but also bread and turkey for her six children, of whom one was an idiot and another a cripple. Whilst such gifts, where they occurred, presumably improved employer–employee relations, they were nevertheless merely intended as acts of Christian charity and not as incentives for harder and more innovative work.[102]

The only practice that was used by some British mills to extract innovation and greater effort from their workforce was that of tonnage. Effectively this was a bounty paid on every ton of perfect paper produced over and above a certain minimum level of output that had previously been specified by the mill. When it was operated fairly, and it was possible for all workers to benefit from it, it could have desirable effects.[103] The problem was that the practice was open to abuse from employers. Indeed, many workers clearly considered tonnage a major grievance and when it was discussed at the 1895 annual general meeting of the NUPMW, it was unanimously resolved that steps should be taken to get the 'unfair and distasteful system abolished and a fair day's pay for

---

[100] *RC Labour*, p. 761. The only example that I have come across was Bertrams, the paper-machine makers, which awarded prizes for suggestions that resulted in improvements in workshop procedures and machinery. *PMC*, 10 September 1891, p. 330. Some firms waited until after the Second World War to introduce bonus schemes. Watson, *Last Mill*, p. 66.

[101] Board of Trade, *Report by D. F. Schloss on Profit-sharing*, p. 80; *Report on Profit-sharing and Labour Co-partnership in the United Kingdom*, p. 15.

[102] Hollingworth Collection, U1999 B8/1. As another example, Frank Lloyd donated £1,000 and the company £500 to the office's Sick and Burial Fund. *PMC*, 10 July 1896, p. 333.

[103] Ketelbey, *Tullis Russell*, pp. 152–3.

a fair day's work substituted'.[104] The two main problems with tonnage were that, firstly, the amount of output above which it was necessary to go in order to qualify for tonnage was variable and subject to increase usually as the workers were drawing near to it; and secondly, that even when it was being paid out, not all workers were party to the practice.[105] The benefits almost always only accrued to the skilled labour in the mill. The practice was also frequently used by employers to keep organised labour out of the mills. An illustration of some of these factors can be seen in the development of the tonnage system in the first decade of the twentieth century at Cannon and Clapperton Paper Mill in Sandford-on-Thames, Oxfordshire. The essential features are shown in Table 8.5. The steady creeping up of the amount needed to receive tonnage is apparent as is the difference in the level paid according to occupation. It is also worthy of note that large sections of the workforce, such as finishers and labourers, were excluded totally from the practice. After 1903 not only was the tonnage minimum steadily raised, but tonnage rates also fell or disappeared. The beaterman's tonnage rate, for example, fell to 1s, while the extension of the practice to the boiling house in January of 1903 was withdrawn three months later. One other interesting aspect, written and underlined in red ink in the wage book, was the instruction that no union or suspected union men were to receive tonnage at all.[106]

A further factor preventing workmen from sharing their ideas with management were the foremen, many of whom acted as very real obstacles to communication between the boss and the shopfloor. Such opposition from foremen to those who had ideas was in part due to jealousy and an unquestioning faith in old rule-of-thumb methods, but it was also in part due to the fact that many employers themselves actually encouraged that type of attitude in their employees, especially their foremen. Certainly the *Paper Trade Review* appears to have condoned such an attitude:

A foreman may, after length of service, know more about the business than the mill owners, but that is not what the firm engaged him for. There is a steady demand for practical foremen and helpers who have learned to do what they have been told, and not consult their own opinion in the matter. An employé that learns to obey, has learned one of the first elements of success.[107]

[104] NUPMW Reports, 41T/BOT, 1895 Report, p. 3. Also claimed in *RC Labour*, p. 509.
[105] NUPMW Reports, 41T/BOT, 1891 Report, p. 330; *RC Labour*, pp. 509, 755.
[106] Oxford County Record Office: Cannon and Clapperton Paper Mill Records, Ca/Cl III/i/1–3.
[107] *PTR*, 14 December 1883, p. 275. See also *PMC*, 10 March 1903, p. 101; Bundock, *Story*, p. 373.

Table 8.5 *Tonnage scales and rates in Cannon and Clapperton Paper Mill, 1902, 1905, 1907 (in minimum tons necessary)*

Occupation	Rate	1902	1904	1907
Machineman	1/3	25	50	60
Machine assistant	6d	25	50	60
Beaterman	1/3	25	50	60
Assistant beaterman	9d	25	50	60
Stoker	3d	25	50	60
Cutterman	6d		50	60
Calenderman	6d			60

Source: OCRO: Cannon and Clapperton Paper Mill, Ca/Cl III/i/1–3.

This may be so, but a foreman who obeys an impractical employer may find the mill – and ultimately his job – rapidly going to ruin. A firm that silences the innovative amongst its numbers sacrifices a vital source of its own longevity.

It can be said then, that as a result of its attitudes to labour relations in the late Victorian and Edwardian era Britain's paper industry, at least relative to the United States, neglected an important source of technological accumulation. As a consequence, as it entered the twentieth century, America's workers were better able and more willing to contribute to the further economic and technological development of the industry.

Labour productivity in Britain and America, 1860–1913: summary and conclusions

In chapters 4–7 we have sought an explanation for the persistent two-to-one gap in the level of labour productivity that existed between the American and British paper industries in the late Victorian and Edwardian period. By utilising what data there are on output, machines, and employment, we have been able to identify differences in the composition of this labour productivity differential at different times over the period studied. Prior to 1890 it appears that all of this gap can be accounted for by the labour input component of the labour productivity identity alone. In effect that amounts to saying that British paper manufacturers employed more labour per machine or per mill than American manufacturers. In following up this interesting difference between the two nations, it was found that this, in turn, could be attributed to two factors: choice of raw material and labour-saving machinery.

In chapter 7 it was demonstrated that in Britain papermakers used raw materials which required considerably greater input of labour than the materials chosen by their American counterparts. However, despite the fact that this lowered British labour productivity *vis-à-vis* American, this choice, as we saw in chapter 4, was a perfectly reasonable one for them to make. In addition to raw materials another factor influencing comparative labour productivity was the more common employment in the United States of labour-saving devices. This difference was shown to be related to the relative costs of labour, machinery, and energy in the respective countries, and thus did not have its origin in any failing on the part of the British manufacturer. It should be noted that no technological difference or labour pressure for overmanning can be detected in either country in this period.

After 1890 the nature of the labour productivity gap and its explanation change. From that date the factors that were crucial in the preceding period diminish in importance, and new ones emerge into prominence. Our data and the observations of contemporaries alike suggest that America moved ahead of Britain technologically in the last three decades before the First World War. In this chapter it was argued that the faster replacement and acquisition rates of paper-machines in the United States contributed to the surge in its average technological sophistication in this later period; a factor that was largely attributable to the greater scale of the American market. Demand, however, could only account for at most a third of the technological gap between the two countries in this period, and moreover, leaves America's superior generation of innovation unexplained. In an analysis that was based on the actual nature of technological changes in the nineteenth-century paper industry, it was argued that these could be explained by the environment that each industry created for that type of technological change to take place in. More specifically, the seemingly greater alacrity of American producers to co-operate on matters pertaining to the industry, create research and educational institutions, train, reward, and maintain cordial relations with their workers, as well as to expand and vertically integrate their production, was seen as being integral in comprehending America's technological success after 1890. Once again the restrictive practices of unions, industrial structure, and technological differences as explanations for the labour productivity gap after 1890 were found wanting.

Taken together these findings and their effects can be captured and summarised in three diagrams. The first of these diagrams, Figure 8.1, shows the effect of raw material choice on labour input in the rag

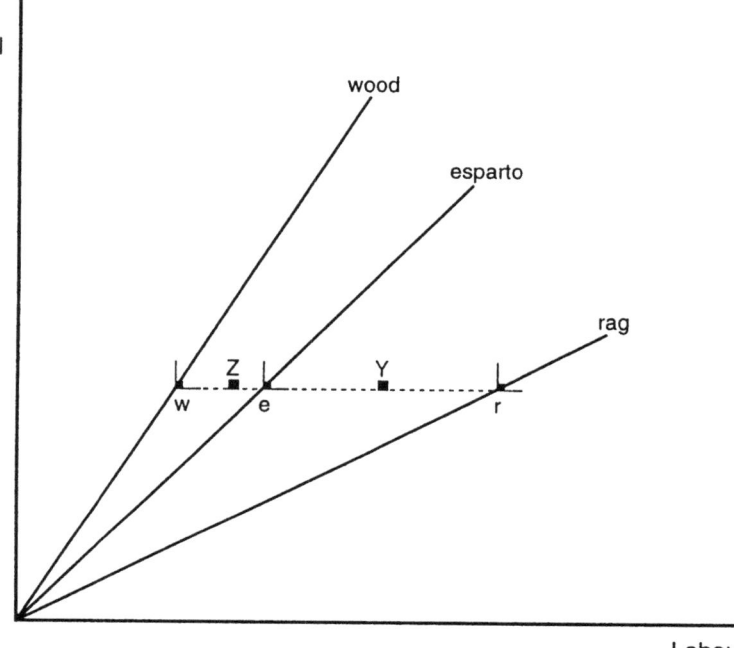

8.1   Raw material choice and labour input

room.[108] In Figure 8.1 the amount of labour needed to make x amount of pulp from the same quantity of wood, esparto, and rag is given. As each material was in essence a different technology,[109] each is drawn with its own individual isoquant. The labour requirements of each material are represented by rays out of the origin. As these are fixed technological relationships, using less labour is impossible, while using more adds nothing to output. Moreover, from what is known of

---

[108] Despite its similarity to linear programming, Figure 8.1 does not concern itself with the actual choice of raw materials.

[109] In all three diagrams a sharp distinction is drawn between technologies and techniques. A *technology* is defined as a body of knowledge which has direct application to an industrial pursuit. This corpus of knowledge in turn embodies a number of different ways of achieving a desired goal, *techniques*, all of which employ the same basic technological principles. Technological change in this case becomes the process by which new knowledge is brought about, while technical change is simply the movement from one technique to another that draws on the same knowledge, and hence requires the development of no new technology. For example, the difference between making pulp from wood and esparto is a technological one, because different bodies of knowledge are involved, while making wood pulp with more or less capital is a technical one, as the same fundamental knowledge is applied. See J. Elster, *Explaining Technical Change* (Cambridge, 1983), p. 94.

nineteenth-century papermaking technology, no substitution between labour and any of the raw materials is possible. That is to say, using more labour and less raw material will always result in a reduction of output. Bearing this in mind, the most technologically and economically efficient way of producing x amount of pulp with wood is at w, with esparto at e, and rag at r. The linear combination of these alternative materials determines the available process frontier, since firms could use a combination of these materials.

In chapters 4 and 7 it was shown that between 1860 and 1880 American papermakers used more wood than the British who instead relied much more on rag and esparto. American manufacturers' greater familiarity with wood pulp in the 1860s and 1870s stemmed from the Civil War: in many ways an accident of history for the paper trade, which expedited the material's introduction into the industry and which given the localised nature of innovation afforded the Americans an initial advantage in its use. In the late 1860s the American industry thus stood at Z, where, although wood and rag is used, wood forms a significant part of the raw materials chosen.[110]

In the Britain of the 1860s, however, the practical know-how as well as the expertise needed to make wood-pulp paper had not been sufficiently worked out to permit the material's widespread introduction. As we saw from chapter 4, this was not due to lack of effort. Firms like Alexander Cowan and Son certainly devoted a lot of time and money to experiment with wood. Nor was it due to ignorance of the possibilities of using wood and straw – these had been known since Koops' experiments at the beginning of the century. The problem facing the British, as well as the American, papermaker was to devise a process that was not only practical, but commercially viable. Prior to the arrival of the research laboratory, this could only be done by trial and error in the mill. Lacking the stimulus the Civil War gave its competitors in America, British producers by 1870 had made less progress in this pursuit. They had, however, made significant headway with the use of esparto grass. Thus, in Figure 8.1 Britain begins at Y, where a mixture of rag and esparto is being used. This combination of raw materials needed more labour to produce x amount of pulp than did the American.

British producers made this choice even though, as chapter 4 demonstrated, it would have been cheaper to use wood exclusively. In fact,

[110] Z's location between the wood and esparto rays should not be taken to imply that esparto is being used in the US. Its location stems from the fact that a linear combination of rag and wood, where wood is much more important than rag, would place the US industry much closer to the wood, and hence the esparto, ray than the rag.

equipped with full and practical knowledge of the use of wood, we would expect both British and American producers to produce w at all prices, except when labour is free. It is important to note that w represents a theoretical extreme, which given the state of knowledge of wood-pulp technology in the late 1860s could not be practically attained.

For simplicity, Figure 8.1 has also assumed a capital intensity in pulp production common to both countries.[111] As we have seen American methods were likely to be significantly more capital intensive than British methods. Relaxing this assumption and increasing the amount of capital used would *ceteris paribus* reduce the amount of labour needed per unit of each material used. In other words, it would cause the isoquant associated with each raw material to shift leftwards. The effect of greater capital intensity then would place the American industry further left along the available process frontier; a process that would also broaden the labour gap with Britain.

Knowledge rarely remains static. As more reliable information on making paper from wood became available in Britain and as producers themselves through trial and error became more familiar and competent with wood, they began to use increasingly more of it in their pulp; a process which in Figure 8.1 moved the industry leftward along the available process frontier towards Z. By the end of the century this point had been reached.

The preparation of the pulp, of course, was only one part of the production process. What happened in the machine room also affected overall labour productivity. In Figure 8.2 different ways of turning that pulp into paper with varying proportions of capital (K) and labour (L) are depicted. Q1, Q2, Q3 are three isoquants representing all the combinations of capital and labour possible of producing the output, Q. The closer the isoquant to the origin, the more advanced the technology employed. For example, in Figure 8.2 Britain is on a lower isoquant, because, as we saw in chapter 5, it was more technologically advanced in this period. Moreover, the further left one goes along an isoquant, the more capital-intensive the techniques used become. c1, c2 are isocost lines, which reflect relative factor costs in each country, while the rays out of the origin labelled α and β represent particular degrees of capital intensity. The model of localised innovation through learning expounded in this book would suggest that technological progress in each country ought to proceed around these rays.

---

[111] Figure 8.1 also assumes that the pulp conversion rates for each of the materials were the same. While this is not strictly true, any differences in these rates that could have been attained in nineteenth-century papermaking were unlikely to have been significant enough to alter the conclusions of Figure 8.1.

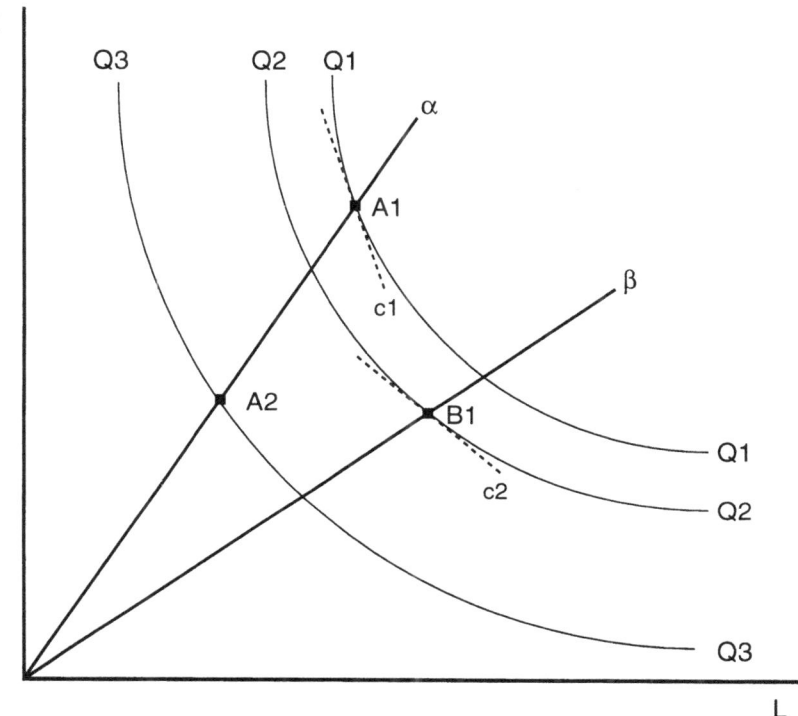

8.2   Technical choice in American and British machine rooms, 1861–1913

Our analysis begins in the late 1860s with the US industry at A1 and the British at B1. America's relative factor prices (c1) favour capital-intensive production, while Britain's encourage more labour-intensive production at B1 on isocost line c2. One effect of this choice was to place British labour productivity, despite its technological lead, below American. Of course, as this diagram does not make any allowance for America's lower labour requirements in the preparatory stage of production between 1860 and 1890, it understates the magnitude of the overall labour productivity differential in papermaking in this period. After 1890 America's faster rate of technological accumulation and acquisition of new machines pushed it ahead of Britain technologically as well. Assuming for the sake of simplicity that there had been no innovation in either country before then,[112] production in America

[112] This admittedly unrealistic assumption is employed purely to simplify exposition. In no way does the analysis depend upon its veracity.

would have as a result shifted down to A2 on isoquant Q3. This would have widened the labour productivity gap in the machine room between the two nations.

The information contained in Figures 8.1 and 8.2 is compressed into Figure 8.3 where the whole process of papermaking from the preparation of the pulp to the actual making of the paper is considered. In structure and appearance Figure 8.3 is almost identical to Figure 8.2, though differs from it in that Figure 8.3 assumes that the overall production functions in Britain and America in 1860 were not the same.[113] This is reasonable because, as we have seen, the amount of labour needed to produce a given amount of paper from unprocessed raw materials even with the same degree of capital intensity varies with the material utilised. For this reason, qb, Britain's initial isoquant is to the right of qa, the American's, despite the technological lead in machine production it held at this time (without which it would be further to the right).

In the late 1860s, given its relative factor prices (pb), the British industry operates at D, while US production takes place at the more capital intensive point E. As a result labour productivity in Britain (1/x4) is half of that in America (1/x2). By 1890 the alteration in the composition of Britain's raw material usage brings about a convergence of production technology, so that Britain now produces at F on isoquant qa. This change alone ought to have significantly narrowed the magnitude of the labour productivity gap (1/x3 − 1/x2) between the two nations. However, by this stage American paper-machine technology outstripped British, shifting the industry down to G on isoquant q3; an event that allowed the two-to-one gap (1/x3 − 1/x1) to be maintained.

One implication of this analysis of the labour productivity differential is that it indicates that the basis of America's lead changed over time; a fact that if generally true would weaken attempts to account for the overall productivity gap in manufacturing with one simple, common explanation. In particular, the evidence presented in chapter 6 is very critical of those theories that pinpoint the inimical practices of British unions as the main culprits for the nation's laggardness. Indeed, one cannot even find much support in the paper trade for the notion of a distinct and persistent British 'system' of manufacturing dominated by batch production, piece-rates, and craft-based unions. It needs to be made

---

[113] Although a fundamental production function embracing all raw materials could have been hypothesised, this, of course, does not mean that at any time all points on its isoquants could have been reached. Instead, for relative ease of exposition, Figure 8.3 has assumed a separate production function for each state of knowledge concerning the different cellulose-extracting technologies.

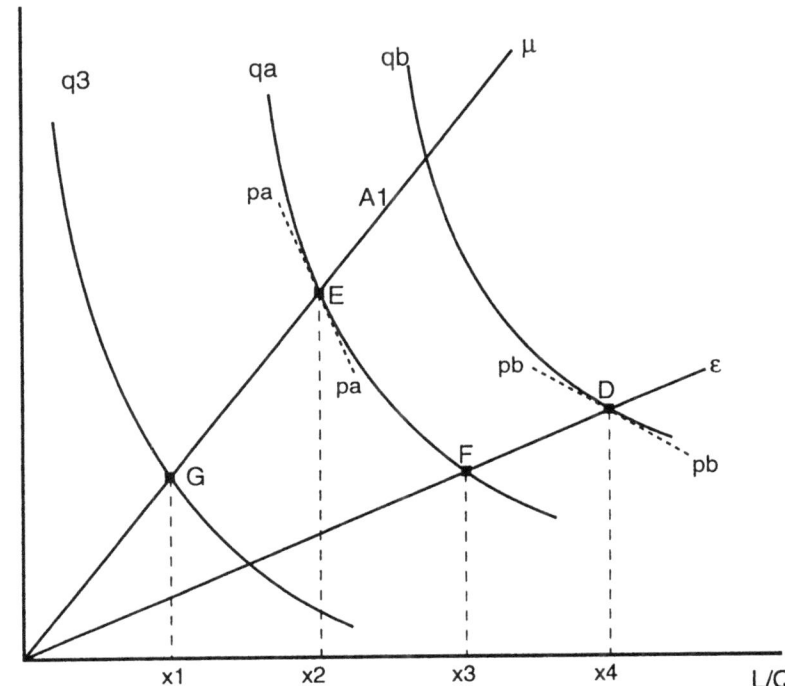

8.3 Technical choice in the American and British paper industries, 1861–1913

clear here that my objection is to the stereotyping of the British paper industry in such a way, not to the contention that its nature and shape reflected its distinct environment. After all, America's use of labour-saving devices and greater throughput, especially after 1890, illustrate the importance of the pattern of demand and resource and factor endowments. My point then is that it is wrong simply to assume from this that the main aim of British papermakers was always to utilise labour – especially skilled labour – intensive production technologies, or that the industry was necessarily craft based.

Moreover, it is equally wrong to assume that any differences in the nature of production in American and British papermaking that did exist were of a constant and immutable character. The paper industry provides no evidence of permanent lock-in. On the contrary, it illustrates that initial paths can be changed. In the decades following 1860, although machine technology in both countries was similar, cellulose-extracting technology differed, with British producers employing the more labour-intensive method. However, as wood became popular later in the century, British and American cellulose-extracting technology steadily converged, so that by 1900 the machine and cellulose-extracting technologies in both countries were almost identical. By this time, however, the faster rate of technological change experienced in America enabled it to maintain its two-to-one productivity lead.

The analyses in chapters 7 and 8 also confirm Broadberry's assertion that technological leadership need not imply productivity leadership.[114] As we have seen, although Britain led America technologically between 1860 and 1890, its labour productivity always remained half that of the American industry. The findings are also broadly consistent with the view that America's post-bellum economic supremacy was largely tied to its resource endowment, technological choice, and mass markets. The faster rate of innovation it achieved after 1890 was also an important factor. America's bounteous resource endowment not only provided its paper manufacturers with raw materials that enabled them to cut their labour requirements dramatically, but also induced them to adopt more capital-intensive production methods. The paper industry lends support as well to those who have attributed America's productivity lead in manufacturing to its larger demand. This augmented productivity not only by promising American producers a faster rate of embodied technological change, but also by initiating structural changes that enhanced throughput and the rate of innovation derived from experience. Opened up by the railroad and protected by tariff walls, the potential of America's massive market increasingly galvanised its papermakers, especially from the 1890s, to expand, specialise, and vertically integrate.

As for the question of failure, the analysis of chapters 6 and 7 has suggested very strongly that prior to 1890 the mere existence of such a gap alone cannot be regarded as an indictment of the British paper industry, for, as has been argued, the decisions that created it were made on purely rational bases. This is not to say that all British paper manufacturers performed adequately, but it does say that for the industry as a whole no such failure can be discerned. After 1890, however, some evidence of failure starts to emerge. Although the faster growth of demand in America had some part to play in its faster technological accumulation, it is hard to escape the conclusion that a growing lack of entrepreneurial vigour was also creeping into British papermaking by the Edwardian period. This loss of vigour emerged not

---

[114] Broadberry, 'Technological leadership', p. 292.

so much because the attitudes of British papermakers had changed so markedly since the mid-century, but because the business culture, in which they had been nurtured and in which they continued to nurture their successors, proved itself to be less adaptable in new contexts and to new ideas. It was a business culture that regarded traditional hierarchies and relationships in the mill as not only immutable, but virtuous, and which saw the shopfloor more as a potential source of conflict than of co-operation. It was also a culture that judged practical, on-the-job experience as the only useful form of learning and contact with competitors beyond the sociable as unnecessary and potentially deleterious to long-term prosperity. In general it was an outlook that aroused in papermakers misgivings about, if not actual contempt for, ideas and concepts that challenged the long-held beliefs that had served them and their fathers well. These misgivings were most clearly reflected in the reluctance they showed to technical training schemes, industry co-operation, and productivity related remuneration policies. As yet, this aversion to albeit still unproven innovations hardly dealt the industry an immediately fatal blow, and indeed to many at the time it appeared fairly innocuous, but in it was the beginnings of an enduring conservatism that was to dog the fortunes of later generations of British papermakers.[115] Of course, this is not to claim that all was ideal in the United States. There too, individual papermakers displayed the same degree of resistance to new practices that most British manufacturers did. The difference was that in America a relatively larger number of papermakers seemed to have shown a growing awareness that innovation and productivity growth are not automatic or costless by-products of productive activity – as standard learning-by-doing accounts suggest – and that firms either independently or collectively in associations did have a role to play in providing the right environment for innovation.

[115] See, for example, C. F. Carter and B. R. Williams, *Industry and Technical Progress* (London, 1957), ch. 5.

9 Free trade and paper

Although in terms of productivity levels the United States led the world throughout the latter half of the nineteenth century, as an exporter of paper America was to have little impact on world markets until the very end of the century. To a large extent this was due to the size of America's domestic market and high transportation costs, which together diminished the need for, and ability of, Americans to export. In terms of direct competition in the marketplace then, Britain's main rival in this period was Germany, not America. It was a rivalry that extended to both Britain's external, as well as internal paper markets, and which was often tinged with antagonism. Of course, Germany was no marginal producer of paper. After America it was the largest paper producing country in the world, and in 1913 its share of world paper and board production stood at around 20 per cent.[1]

The success of German and other continental paper manufacturers in their own home market was a constant and worrying feature of the British papermaker's experience after 1861. To many observers such success represented an indictment of Britain's commercial policies, which gave foreigners free access to the British home market without in turn securing similar admission to foreign markets for its own producers. In particular, Germany evoked vehement criticism in Britain for its unwillingness to play cricket in the matter of fair trade; though, of course, Germany was by no means the only country in the latter half of the nineteenth century to be censured by British public opinion for its abrogation of the British sense of fair play. In this chapter we determine how justified such criticisms were by considering the consequences of Britain's free trade in paper on the development of its paper industry. In particular, the trade in paper between Britain and Germany between 1861 and 1914 is examined.

---

[1] A. Hülbrock, 'Organisation und Preisgestaltung auf dem deutschen Papiermarkt unter besonderer Berücksichtigung der Gegenwart', Ph.D. thesis, University of Frankfurt-am-Main (1927), p. 41.

Stylised facts

It will be recalled from chapter 3 that, despite some fluctuation due to their different positions in the trade cycle, British labour productivity levels were at all times above those found in Germany. Although at one point British labour was as much as one-and-a-half times as productive as German labour, the more usual magnitude of the labour productivity gap between the two nations was of the order of 10–15 per cent. Given the factor endowment that prevailed in nineteenth-century Germany, this labour productivity differential is hardly surprising. As was well-known to contemporaries, the wages paid in German mills were on average about 75 per cent of those found in British mills. According to an American report of 1909, for example, for a twelve-hour day an average German machine tender was paid the equivalent of $1.80 (American), a beaterman 90 cents, and an unskilled hand 75 cents; while their counterparts in Britain received for the same amount of work $2.75, $1.20, and 90 cents respectively.[2] With such a wage differential between the two countries, one could reasonably expect *ceteris paribus* German papermakers to adopt more labour-intensive methods of production than British papermakers. This indeed appears to have been the case. As the German correspondent of one trade journal reported in 1884, the average German paper-mill usually required as much as three times as many workers as its equivalent in Britain did.[3] As German paper technology was also no better than British at this time, this degree of difference in manning requirements was bound to make labour productivity lower in Germany.

This, of course, does not mean that German paper machinery was in any way deficient. Especially from the 1880s a thriving German export trade in paper machinery developed. As a contributor to a work commemorating the achievements of Kaiser Wilhelm's twenty-five years of reign proudly boasted: 'More so than any other, the German paper-machine has not only driven foreign machines from the German market, but is much sought-after overseas. As such a machine costs at least 300,000 Marks, its export is an economic success.'[4]

Amongst their customers German paper-machine makers could certainly count numerous British papermakers, such as Frederick Pratt Barlow, who bought German machines not because of the superior technology embodied in them, but because of their sturdiness, reliability,

---

[2] *Pulp and Paper*, vol. V, p. 3031.   [3] *PMC*, 10 April 1884, p. 95.
[4] Ing. Heinel, 'Die Maschinen-Industrie', in *Deutschland unter Kaiser Wilhelm II* (Berlin, 1914), vol. II, p. 554.

and above all cheapness.[5] Confirmation of Britain's technological lead is found in the average machine widths operating in Germany and Britain between 1897 and 1914. As increasing widths represented one of the two key areas of mechanical improvement in the industry and involved a complicated series of adjustments and innovations, the average width of paper-machines in a country gives us a rough estimate of the technological sophistication in that industry at that time. Since British paper-machines were on average about 15 per cent wider than German machines, this strongly suggests that the British industry must have *ceteris paribus* employed more advanced machinery.[6]

Despite this, a growing German penetration of British markets was characteristic of Britain's post-1860 experience. Williams in his famous invective against German imports was painfully aware of this, warning his readers that not only was the paper they used most probably made in Germany, but even 'the material of your favourite (patriotic) newspaper'.[7] Tables 9.1 and 9.2 depict this invasion. The 1880s and 1890s appear to be the decades in which German impact on the market reached its apogee. If anything, these figures underestimate the share of imports coming from Germany, as some German paper must have been shipped to Britain via Holland.[8]

This is also true for the figures before 1875, for only paper exported from a port lying in what was to become Germany has been considered German in our calculations for those years. After reaching its peak at about 30 per cent (38.6 per cent if all German exports to Holland ended up in Britain) in 1885, Germany's share of paper imports to Britain began to fall, even though, with the exception of the freak year of 1905 accounted for by a large upswing in domestic demand,[9] the absolute quantity of paper imported from Germany continued to rise steadily. The fall in import share is largely attributable to the growing competition of Scandinavian and North American producers who at this time were beginning to make inroads into the British and continental markets. In 1885, for example, North American (US and Canada) paper accounted for less than 1 per cent and Scandinavian (Sweden and Norway) about 16 per cent of the total paper and board

---

[5] *RC Depression*, p. 598. See also C. Franck, 'Modern paper making machinery', *German Export Review* 15 (Sept. 1901), 103.

[6] For example, in 1900 the mean widths of machines in Britain and Germany were 77.3 and 66.2 inches respectively. Data come from *Paper Mill Directory of the World*.

[7] E. E. Williams, *Made in Germany* (London, 1896), pp. 10–11.

[8] This at least was the impression of George Chater. *RC Depression*, p. 593.

[9] J. Geuenich, *Geschichte der Papierindustrie im Düren-Jülicher Wirtschaftsraum* (Düren, 1959), p. 66.

Table 9.1 *German exports of paper and board to the UK, 1865–1913 (including German exports to Holland in parentheses)*

Year	Total imports of German paper and board (cwt)	German paper and board imports as a percentage of all paper and board imported into Britain	German paper and board exports to Britain as a percentage of all German paper and board exports
1860	1,249	4.4	
1865	5,956	2.5	
1875		17.2[a]	
1885	426,369	30.8 (38.6)	41.8[b]
1895	746,528	20.4 (24.9)	40.7
1905	543,284	6.8 (8.7)	26.5
1913	1,708,312	13.3 (16.5)	28.0

Notes: [a] Revenue share. [b] Excluding exports to other German states.
Sources: Annual Statement of the Trade of the United Kingdom, and *Statistiches Jahrbuch für das deutsche Reich.*

imports into Britain in that year. By 1913 these had risen respectively to 9.7 and 36.3 per cent.[10]

The British market was the single most important market for German exporters, with almost half of all German exports ending up in Britain. In particular years such as during the Boer War, in which newspaper circulation grew enormously, German paper reached even greater proportions. In 1900, for example, the import of unprinted German paper used by newspapers soared to 473,093 cwt, and German commentators wrote of a heyday for the German industry where they could dispose of all they wanted: 'Englische Einkäufer boten in Deutschland jeden Preis, um nur Ware zu erhalten und nahmen jede Mengen ab. Der Papiermarket war in lebhafte Aufregung verfallen.'[11] In the first two decades of the twentieth century this proportion fell, as German producers, forced out of the British market by Swedish, Norwegian, American, and Canadian papers, increasingly sought markets in Asia, Australasia, and South America.[12] Indeed, by the outbreak of the First

---

[10] All trade figures in this chapter, except where otherwise stated, come from *Annual Statement of the Trade and Navigation of the United Kingdom with Foreign Countries and British Possessions* (London, 1860– 1913), and *Statistiches Jahrbuch für das deutsche Reich* (Berlin, 1880–1913).

[11] F. Demuth, 'Die Papierfabrikation', *Schriften des Vereins für Sozialpolitik* 107 (1903), 203.

[12] Not without success. In 1895 17.2 per cent of all German exports of printing and writing paper went to South America. By 1913 this figure had risen to 33.5 per cent.

Table 9.2 *German exports of printing and packing paper to the UK, 1865–1913 (including German exports to Holland in parentheses)*

Year	German imports of printing and packing paper (cwt)	German imports of printing and packing paper as a percentage of all printing and packing paper imports	German exports of printing and packing paper to Britain as a percentage of all German exports of printing and packing paper
1865	5,732	3.2	
1875		13.2^a	
1885	266,288	37.4	
1895	460,713	22.1 (27.6)	43.2
1905	398,457	8.2 (10.0)	27.2
1913	1,367,033	17.7 (21.6)	31.6

Note: ^{a} Revenue share.
Sources: As for Table 9.1.

World War the Argentine Republic had become as big an importer of German printing paper as Britain. These were all traditional British markets – with over 70 per cent of all British paper and board exports consistently going to its colonies and possessions – so that the relative decline in German penetration of Britain's home markets did not spell an end of the threat from Germany. This was also the story of the printing and packing paper sector of the industry. The decline in importance of German exports in this sector in the twentieth century is explained by the fact that it was precisely in this sector, which could best utilise cheap wood pulp, that foreign competition for the German manufacturer was most acute. Partially as a response to this, the German industry diversified into other more specialised paper and paper-related products such as photographic, fancy, and coloured paper, as well as postcards. These in turn found a ready market in Britain.[13]

In the late Victorian and Edwardian period German competition thus came to be an important element in the British papermaker's experience. In light of Britain's not too unfavourable productivity and technological performance, this strikes one as somewhat puzzling; at least in terms of

For the purposes of comparison with Table 9.2, 19.7 per cent of German printing and packing paper exports in 1913 went to South America.

[13] C. Buchheim, *Deutsche Gewerbeexporte nach England in der zweiten Hälfte des 19. Jahrhunderts* (Ostfindern, 1982), *passim.* Buchheim points out (p. 105) that even much of the paraphernalia associated with Queen Victoria's diamond jubilee, official and otherwise, was made of German paper products. See also *German Export Review* 15 (June 1901), 76–8.

the sheer scale of the German penetration of the British markets. Certainly, the same cannot be said in reverse: Britain's exports to Germany represented a meagre 3 per cent of the total of all imports to that country. In a sense then, the suspicion is that, given the productivity and technology figures, the German paper industry was doing just too well.

British explanations

At least to the British papermaker the origins of this puzzle were obvious. The gravamen to him was Britain's seemingly blind adherence to the principles of free trade. To be precise, it was not the principle itself that so gravely offended, but the practice of free trade, for it was felt that what was taking place at the time all around them was hardly 'free', and certainly not fair. 'Perfectly free and fair trade would, however, satisfy the desire of every British manufacturer who only wishes fair play and no favour', proclaimed the papermaker Alexander Annandale, who then went on to complain that the reality was anything but fair.[14] John Evans, president of the Paper Makers Association and Fellow of the Royal Society, lamented that 'our system of free trade is gradually acting so as to increase protection abroad, and to render that protection not simply defensive to those foreign countries, but really to give them the power of aggression here'.[15] Faced by steep foreign tariffs and belligerence to their paper abroad, British papermakers saw themselves in a tight, no-win situation. John Evans eloquently expressed the papermaker's dilemma to the Royal Commissioners in 1886:

In order to meet the foreign competition, the great question with the English manufacturers was how to cheapen their production; and, partly owing to the increased demand, owing to the abolition of the duty, and partly owing to the desire to reduce costs most of the paper-mills increased their power of production very largely; and, as a consequence, by increasing the production they were enabled to reduce their standing charges, and so reduce the general cost of their manufacture, which to some extent for a time enabled them to compete with foreign countries. But of late years foreign manufacturers have been going upon the same task, and have largely increased their mills; and now I think that on the whole the production is in excess of the demand, and the great question that one has to consider is the way in which one can dispose of the surplus products, that is to say, the excess of our manufacturers over what would be the ordinary demand in order to keep a mill cheaply employed. If we attempt to export our surplus products to any continental country, or to the United

[14] Reply of Alexander Annandale to the circular addressed to the principal commercial associations. *RC Depression*, appendix C, pp. 428–9.
[15] *Ibid.*, p. 585.

States of America, we are met with heavy duties; whereas the continental manufacturers have here a ready market for all their surplus products, and those surplus products are sold below the ordinary rate which is maintained within the protected countries; ... surplus production, which the manufacturers can afford to sell somewhat below the ordinary costs of their production, and yet realise a profit.[16]

Each of the other witnesses from the trade, who were called before the Royal Commission, as well as those who appeared before Chamberlain's Tariff Commission twenty years later, concurred with the gist of Evans' assessment of the situation and identified German producers as the main culprits.[17] One British boardmaker, for example, told the Tariff Commission that a German-made cardboard, which cost £6 per ton to produce in Britain and which was sold for £7 in Germany, could be had in Britain for only £5; a feat that he and others maintained was only made possible by the fact that German producers were prepared to sell their surpluses at a loss.[18] Of course, American and other continental manufacturers were also accused of employing similar tactics, though Germany always seemed to be the primary target of the British papermakers' ire.[19] Not only was Britain's home market vulnerable to the dumping of surpluses, but so were its main export markets, the colonies: 'We are hardly able to hold our own, because orders are taken and paper is supplied at prices which are unremunerative to the English manufacturer', was the complaint of one papermaker.[20] Nor was there much opportunity for British producers to retaliate in kind by inundating German markets with their own excess production, as these were heavily protected by tariffs. In the 1880s, for example, these protective duties stood at the equivalent of a halfpenny per pound of

[16] *Ibid.*, p. 586.
[17] See the comments of Barlow, Chater, Annandale, the Scottish Paper Makers Association, and the reports of various chambers of commerce in districts with large paper-mills (e.g. Barnsley district) in *ibid.*; Tariff Collection, TC7 28/2 B(13), and TC3 1/89 p. 4. Trade journals also carried numerous instances. For example, a papermaker in 1886, who had just visited a number of mills in Germany, noted that 'they are satisfied to undersell us in our markets at a shade above the costs of production, and more often at a *loss*, yet they make their own people pay a price which will recoup them for their losses by this pernicious system of competition'. *PMC*, 10 August 1886, p. 263. Other examples include *PTR*, 1 June 1863, pp. 121–2; *PMC*, 10 February 1887, p. 39, and 10 November 1903, p. 411.
[18] Tariff Collection, TC7 28/2 File 167 (B)4.
[19] See *ibid.*, TC7 28/2 (B) 3, (B) 20, (D)8. David Cowles, the president of the American Paper and Pulp Association, gave his views on European producers to a Select Committee in America in 1908: 'they are notorious dumpers. I have done that same thing myself and I know what it means. They will dump a whole lot of paper in a market under certain conditions and get rid of it and sell it at a loss.' *Pulp and Paper*, vol. II, p. 910.
[20] *RC Depression*, p. 591.

paper; a figure that amounted to a massive proportion on many varieties of paper. Indeed, it represented about 33 per cent on the cheapest papers that Britain was then exporting to the German Empire and effectively priced such paper out of the market.[21]

Lest it be misunderstood, this manipulation was no accident, but the deliberate policy of many manufacturers. Frederick Pratt Barlow just prior to his appearance before the Royal Commission visited seventeen mills on the Rhine and in Saxony and Bohemia as a member of a delegation of mainly Scottish papermakers who sought the source of German competitiveness. There they were greeted warmly and the mills they inspected were opened to them, so that they 'were shown everything'. As a result Barlow felt competent enough to speak as an expert on the German industry. He told the Royal Commission of one German producer who had explained the secret of Germany's success to him:

They seem to look upon England, as one man expressed to me, as a sort of rubbish heap, on to which he could shoot all his surplus products by getting rid of so much a week at cost price, or a little below. He explained that he was able to keep down the expenses of the mill, and so make a handsome profit on the paper that he sold in his own market, where it could not be disturbed at all by English competition. There was one particular paper which we were shown that was selling in Berlin for 3d., and the same paper was selling in London for 2d., notwithstanding its coming from the furthest end of Saxony, by rail and river, and consequently being transshipped no less than three times. At one mill – I was not at that particular mill – they told my friend with glee that they were making English postcards for the English Government; or at least, for a contractor to the English Government.[22]

Barlow's last point touches on another matter that was a source of much public attention and debate at the time and which represented another arm of the competitive strategy of German and other continental producers. Essentially this involved the producers and their English distributors selling foreign paper as British paper. This was done by foreign producers imitating British watermarks and by local distributors repackaging foreign paper with British labels. Although this was always difficult to prove, its occurrence appeared to be general knowledge within the trade at the time. As a result, consumers, apparently even HMSO, were often unaware that the paper they were purchasing had its origins overseas.[23]

---

[21] *Ibid.*, p. 598. In 1864 French and Belgian rates varied from 10 to 20 per cent, Austrian and German from 20 to 36 per cent, and American 35 to 40 per cent. Ketelbey, *Tullis Russell*, p. 105.

[22] *RC Depression*, p. 597.

[23] *Ibid.*, p. 587; *PTR*, 1 August 1863, p. 156, and 28 November 1884, p. 331; *PMC*, 10 February 1896, p. 66; Tariff Collection, TC7 28/2 File 167 G(3).

British producers were not alone in their condemnation of foreign exploitation of Britain's open door. Trade unionists likewise saw it as among their chief ills. The preamble of the NUPMW's rulebook, written in 1890 at the height of the German invasion, clearly demonstrates the union's anti-foreigner stance:

Further, no effective patriotic instrument of legislation for the defence of an important British and Irish industry, seriously threatened and affected by the hostile action of foreign countries, and handicapped in many ways in competition with imported produce, can be obtained without such a union.[24]

As far as possible, the NUPMW acted to restrict the penetration of foreign paper, especially when achieved by subterfuge. At the Belfast Trades Union Congress in 1893 and subsequent congresses, its general secretary, William Ross, continually raised motions to ensure that foreign-made paper was clearly marked as such, and that all Trades Unions, Trades Councils, and Co-operative societies purchased only British-made paper. This proposed prohibition of foreign paper was later extended to paper used in government offices and contracts.[25] Finally in 1895 at the Cardiff Trades Union Congress Ross succeeded in getting these motions accepted as Congress policy. In addition he got the Congress to adopt the following motion that targeted the foreign producer rather than the local distributor:

That this Congress is of the opinion that in order to prevent foreign paper being imported and palmed off, after changing the outside wrapper, as home manufacture, every sheet of foreign paper imported into this country must have a watermark which can be seen, if the sheet is held up to the light, showing the country it was made in; and every reel of foreign paper must have a similar watermark on at least every square yard. All foreign envelopes, paper bags, and such like must be made of paper which originally had a similar watermark on every square yard.[26]

Of course, it was one thing to get the Trades Union Congress to agree to it, yet another to get anyone to act upon it. The problem clearly continued. In April 1901 the Amalgamated Society of Paper Makers joined the NUPMW's attack on the foreign infiltration of the paper market and issued a joint resolution 'viewing with alarm the state of the paper trade' and calling for the extension of the Merchandise Marks Act to cover all types of paper and board, including newsprint, so that the source of the paper by law would have to be clearly printed on each sheet

[24] Bundock, *Story*, p. 371.
[25] NUPMW Records, 41T/BOT. 1893 and 1894 Reports.
[26] *Ibid.*, 1895 Report, p. 4.

(the first page only for newsprint), and not just on the outside wrapper as was the prevailing custom.[27] Newspaper proprietors were seen as the worst abusers. In 1893 Ross undertook to write to 100 of the nation's leading newspapers asking if they would be prepared to print the words 'Printed on Paper made in the United Kingdom' on the cover of all their publications. Perhaps indicative of the pervasiveness of the practice of using cheap foreign paper, only eighteen replied in the positive.[28] George Chater, before the Royal Commission, also claimed that a number of the great English newspapers contracted directly with foreign manufacturers for their surplus production, although he declined from publicly naming them.[29] In fact, so frustrated was Ross with the behaviour of certain British newspapers on this matter that he rather naively introduced a motion at the Edinburgh Trades Union Congress in 1896, that aimed to ban from the Congress reporters from newspapers and periodicals which could not prove their exclusive use of paper made in the United Kingdom by trade union labour. Needless to say, the Congress quickly buried the motion.[30]

To all concerned these measures were at best palliatives, treating the symptoms rather than the cause of the problem. If genuine free trade could not be attained, then retaliatory and punitive measures to shock the Germans and others back into line was the preferred course of many in the trade. As the Scottish papermaker, Alexander Annandale, forcefully explained:

The tariff of one country should be dovetailed into that of another, according as each individual country shows itself considerate or the reverse. This as a great producing country should always retain the power in her own hands to alter and to regulate her tariff as is most advisable in the interest of her commerce and as she is being treated by other countries. Were this followed out sharply rather than the worship of a mere principle, we should hear of and have fewer protective or prohibitive duties abroad. The rule that has always proven correct in actual warfare holds equally true in fiscal arrangements, viz., that a strong power of retaliation is a corrective.[31]

Free trade, however, remained and with it the disposal of surplus product on the British market. At the turn of the century the practice of dumping became institutionalised in Germany, as what had formerly been the actions of individual producers and the result of temporary agreements between local associations of papermakers became the

[27] Bundock, *Story*, p. 388.   [28] NUPMW Records, 41T/BOT. 1893 Report, p. 5.
[29] Evidence of Chater, *RC Depression*, p. 592.
[30] NUPMW Records, 41T/BOT. 1896 Report, p. 3.
[31] Reply of Annandale, *RC Depression*, pp. 428–9.

prerogative of large cartels and syndicates. This metamorphosis, which is discussed later, introduced little that was new, apart from a wave of panic amongst British papermakers; a fact made evident at the time by the unprecedented publication of two parliamentary papers on the German paper industry.[32]

Alternative explanations for the German industry's success were offered, particularly by those who cherished the principle of free trade. The most interesting of these attributed German success to the fact that Germany, unlike Britain, had been blessed with a resource endowment exceptionally beneficial to the paper trade. Thus, even though German mills were on average less productive than British, they could still outcompete the British by virtue of the much lower input costs they faced. Anecdotal evidence, however, would suggest that the importance of this factor is too easily overestimated. While there was no doubt that wood was in more abundance in Germany and that this was an advantage for the German producer, British papermakers were not convinced that this represented the crucial difference between the two industries. Indeed, Evans expressed the view that the ordinary price of paper in Germany was certainly no lower than in Britain and most probably higher.[33] Frederick Pratt Barlow, who had just returned from Germany where he had made enquiries on this issue, was more certain, confidently reporting that, taking all things into consideration, material costs were comparable in the two nations. If wood and straw had been slightly cheaper in Germany, then by his reckoning this was adequately compensated for by the cheaper coal and chemicals in Britain: a claim also supported by various reports in the trade literature. Moreover, because the Germans on average tended to employ as much as three times 'more labour in their mills, the cost of labour per ton of paper produced was also reputedly similar in both places.[34] Wittigschlager's history of German machine-made paper supports the substance of Barlow's claim by suggesting that the largest component of reduced paper prices in the second half of the nineteenth century came via greater scale, rather than lower wood prices.[35]

---

[32] Report on the German Paper Industry and Export Trade by Consul-General Sir A. Ward, *PP* CXXII (1906), 173–85; and Abstract of the Proceedings of the German Commission on Kartells in *Memoranda, Statistical Tables and Charts Prepared in the Board of Trade with Reference to Various Matters Bearing on British and Foreign Trade and Industrial Conditions* (London, 1905), vol. II, 409–11. Spicer, *Paper Trade*, appendix B gives another example of the contemporary perception of the German industry.

[33] *RC Depression*, p. 587.

[34] *Ibid.*, pp. 597, 599–601; *PMC*, 10 April 1884, p. 95. Another claimed that coal was cheaper in Britain. *PTR*, 1 February 1864, p. 58.

[35] K. Wittigschlager, 'Geschichte der Maschinepapierfabrikation in Bayern rechts des Rheins bis 1914', Ph.D. thesis, University of Frankfurt-am-Main (1930), p. 96.

Further doubt on the singular importance of Germany's putative abundance of wood pulp is cast by the fact that German producers themselves bitterly complained about the difficulty of getting cheap raw materials. In 1913, for example, the Augsburg Chamber of Commerce, reported that only around 40 per cent of the region's wood pulp needs had been supplied that year, resulting in the local mills running on half time with all the associated costs.[36] One reason for this, other than that the demand for wood pulp was running far in excess of its supply, was that the type of collusive behaviour that the papermakers were themselves fond of was increasingly being practised by those in the pulp industry. As early as 1878 the *Verein Deutscher Holzstoffabrikanten*, a wood pulp makers' association, was formed and began organising with varying degrees of success regional cartels. In the 1890s four such cartels were active in the wood-pulp and cellulose industry: the West German, Rhineland, and Westphalia syndicate; the Southern German cartel; a Saxon one; and a fourth that was made up of an assortment of wood-grinding mills from various parts of the country.[37] These influenced the supply and price of the raw material available to the papermaker. As Britain's consul-general in Hamburg noted in his report of 1905: 'According to the view of the *German Paper Manufacturer's Journal*, the prominent question which is at present occupying the minds of the paper manufacturers in Germany is the difficulty of obtaining a sufficient supply of raw material.' This situation was aggravated by the producer of half-stuff who was 'reluctant to dispense of his stocks, as he expects to obtain higher prices when the season of greater business activity commences and the demand for paper begins to increase.'[38]

German explanations

One could perhaps treat all this talk of German dumping with some scepticism had these views been solely the opinions of British paper-makers and commentators. The fact, however, that the German literature is broadly in agreement means that we must give these views serious consideration. Contemporary German commentators were particularly frank about what was going on. Hans Drabsch, employing the martial analogy that appeared quite popular at the time, stated in the most forthright of terms the reasons for Germany's victories in the paper trade:

[36] *Ibid.*, p. 122.
[37] Demuth, 'Papierfabrikation', p. 241; Wittigschlager, 'Geschichte', p. 68.
[38] *Report on the German Paper Industry*, p. 184.

We were compelled to seek the market of the world at a very early date as the cheapness of our products forced [sic] makers to turn out goods *en masse*; of course the supply soon exceeded the home demand, more especially when production began to attain ever larger dimensions owing to the continual improvements introduced into the factories and in the technical plant. Following upon this cause the grim struggle for existence which our industry had to wage against foreign competition in the home markets. This sharpened and increased our capabilities and induced us to carry the war into the enemy's camp and fight him in his own markets in order to find openings for our still growing home production.[39]

This strategy of using the British market as a dumping ground for excess German production was carried out both collectively and individually by German papermakers. Geuenich tells us about the firm of Frederich Heinrich Schoeller of the Düren-Jürlicher region which from the 1860s regularly exported large quantities of its output to England,[40] while a report of the Upper Bavarian Chamber of Commerce declares that, as early as in 1861, the year in which British duties on imported paper were repealed, a crisis of overproduction for the local industry was only averted by the opening up of the English market: 'die bedeutende Erhöhung der Papierproduktion hätte beinahe eine Krise zur Folge gehabt, wäre nicht der englische Markt für die bayerische Papiererzeugung geöffnet worden, nachdem dort der Eingangzoll herabgesetzt wurde'.[41]

The presence or absence of tariffs and other barriers to trade were crucial in determining the destination of Germany's exports. Demuth in his analysis of the German paper industry in 1903 for the *Verein für Sozialpolitik* (Association for Social Policy) explained that Germany's biggest export markets, Great Britain, Belgium, and Holland, were such precisely because they had no or relatively low customs duties.[42] This resulted in some strange outcomes. Bavarian producers, for example, some of whose papers ended up in far-off Britain, were blocked from what would have been their natural markets, Austria and Switzerland, by an imposing array of tariffs.[43]

A collective spirit amongst papermakers and an interest and involvement in the industry by government also greatly facilitated the disposal of surplus production. The government, in particular, played an active role in encouraging the industry's technological development and in keeping freight rates for its products down.[44] British papermakers

---

[39] *German Export Review* 15 (June 1901), 75. See also *Deutschland unter Kaiser Wilhelm II*, Part 3, pp. 676–7.
[40] Geuenich, *Geschichte*, p. 514.   [41] Wittigschlager, 'Geschichte', p. 42.
[42] Demuth, 'Papierfabrikation', 218.   [43] Wittigschlager, 'Geschichte', p. 103.
[44] E. Kirchner, *Die Papierfabrikation in den Ländern der Sektion X der Papiermacherberufgen-*

looked upon the latter with great jealousy, complaining that in 1886 the rates of carriage from Saxony to London were about the same as from Edinburgh to London.[45] The Prussian government also founded an official paper testing institute at Gross Lichterfelde near Berlin. Initially the purpose of this institute, established in 1886, was merely to test all paper furnished to the Prussian government offices, but after a very short time it also came to be used by dealers and private persons of other descriptions (both residing in and outside Germany) who wished to have their paper tested according to the rigorous standards laid down by the institute. The testing at Gross Lichterfelde was with a view to establishing the purity, composition, sizing, tone, and moisture content of the paper, as well as to categorise each type of paper by size, weight, consistency, durability, absorptive powers, resilience, and permeability in regard to light. Its fame was wide. A committee set up by the Society for the Encouragement of the Arts, Manufactures, and Commerce in Britain to investigate the declining quality of paper being produced in the country lauded its achievements.[46] At first the German paper manufacturers, however, were less sure of the institute's benefits, regarding it as an unwelcome interference into their business, but in the course of time they came to acknowledge its salutary effect on their industry.[47] By ensuring quality and enforcing a standardisation of the types and forms of paper manufactured, the institute played an important role in breaking down long-held prejudices against the quality of German paper, a development which greatly enhanced its saleability both domestically and externally. German consuls and technical attachés also helped in this regard, in some instances even helping to negotiate deals for German manufacturers.[48] Such factors were invaluable to an industry that needed to dispose of its production as quickly as possible.

As early as 1850 attempts were made to establish an organisation to represent the interests of papermakers more effectively than was possible with the existing collection of *ad hoc* arrangements thrown together by the various local organisations and chambers of commerce. In that year

ossenschaft: historisch-technologische Skizze (Württemberg, 1911), p. 7. Geuenich, *Geschichte*, pp. 108–28 gives examples of government interest in obtaining the most advanced papermaking machinery for the German paper industry.
[45] Evidence of Barlow, *RC Depression*, p. 597. This is not to say that German manufacturers were satisfied with the freight rates they had to pay. The *Verein* on a number of occasions, such as in 1903, petitioned the *Tarifkommission* to lower the rates further. Wittigschlager, 'Geschichte', p. 104.
[46] Society for the Encouragement of Arts, Manufactures, and Commerce, *Report of the Committee on the Deterioration of Paper* (London, 1898), p. 2.
[47] Spicer, *Paper Trade*, pp. 232–3; *Report on the German Paper Industry*, p. 185; *PMC*, 10 March 1896, pp. 117–19.
[48] *PMC*, 10 March 1886, pp. 75–6.

a loose association of southern German papermakers was formed. Based in Mainz, its *raison d'être* was to find ways and means to remedy the serious situation that was perceived to be confronting the paper industry. A meeting of eighteen manufacturers was convened in Düren and a price rise of 8 to 10 per cent was agreed upon. In addition the meeting determined to present a petition to the various German governments requesting a rise in both the rag export and paper import duties. At the Annual General Meeting of 1852 the association went further proposing an economic programme which would have given them substantial monopolistic power in the paper market. A call was put out to bid farewell once and for all to the pettiness that had hitherto racked the industry. Unfortunately, it was all to no avail, as the squabbling and chiselling that had marred all previous local attempts at price fixing, inevitably returned to undermine this effort as well. Ahead of its time, the Southern German association left one lasting reminder of its existence in the continuation of its organ, the *Centralblatt für die deutsche Papierfabrikation*, which carried on being published until the death of its editor Dr Rudel in 1887 when it was incorporated with the *Wochenblatt für Papierfabrikation*.[49]

For the time being the general prosperity and well-being of the German industry undercut any further move towards greater co-operation. However, with the return of difficulties in the early 1870s, this feeling of confidence disappeared, and sufficient stimulus was provided for a renewed call to go out for some type of organisation to represent the new nation's papermakers. The sources of these difficulties were manifold. Firstly, an acute shortage of raw material was felt. Even though mechanically ground wood pulp had been discovered and was already being used, rag at this time still constituted the papermaker's most important source of pulp. This shortage of rag in the early 1870s had developed out of the removal of all duties on their export in 1872, which had had the effect of increasing the already non-negligible flow of rag overseas where higher prices could be had for such material. On top of that, the post-war boom, that had led to the introduction of new machinery and the creation of new productive capacity, came to a sudden and unexpected end in 1873, leaving the domestic market swamped with paper and prices plummeting. Faced with increasing prices for rags, coal, and labour, as well as falling prices for paper, the German papermaker saw his profit margin steadily eroded. The short-

[49] F. Strenger, 'Die Wirtschaftliche und Soziale Bedeutung der deutschen Papier Herstellenden Industrie im Rahmen der deutschen Volkswirtschaft', Ph.D. thesis, Leopold-Franzens University of Innsbruck (1930), pp. 39–41; Wittigschlager, 'Geschichte', pp. 74–6; and *PTR*, 1 January 1864, p. 1.

term response of many was to expand production so as to lower unit costs and extend the profit margin, though this quick-fix in time only tended to compound the problem of overproduction and unremunerative prices and often ended in insolvency and bankruptcy.[50]

In this climate it was realised by large sections of the industry that if the industry was to remain in anything like its current shape and form, some sort of collective action needed to be taken to treat the ailments confronting it. At a well-attended meeting in Augsburg in December 1871, an attempt was made to re-establish the former Southern German association. This was done and a price hike of at least a Kreuzer per pound was decided upon. A similar course of action was decided upon by a group of Northern papermakers.[51] Subsequent events elsewhere, however, were to render these developments redundant.

In 1872 representatives of both northern and southern producers met and decided that the time had come for a nationwide body, the *Verein deutscher Papierfabrikanten* (Association of German Papermakers), to represent the joint interests of both regions of the empire. The first full meeting of the new association took place in Nürnberg on 14 December of the same year. It was attended by the representatives of 81 different mills, comprising together a grand total of 110 paper-machines and 22 vats: some 10 per cent of the industry. Of these, some 67 mills eventually joined. The significant number of mills outside the association in 1872 for the time being stymied the association's desire to exercise monopolistic power, but the *Verein* continued in other ways to carry out its purpoŠe '*die Interessen der Papierfabrikation zu pflegen und zu fördern*'. Through the close contact of its annual meetings a rich exchange of ideas, lacking in the British trade, took place between its members that not only contributed to instilling a common identity in the industry, but greatly facilitated the spread of new business ideas and innovations throughout it.

Officially the *Verein* viewed itself as the industry's representative in negotiations with the authorities, particularly on matters of law and commercial and trade policy.[52] But the *Verein* also exhibited an interest in transforming itself into some sort of cartel. This desire stemmed from the fundamental dilemma facing the papermaker in these times; *viz.*, a rise in the price of paper could only be achieved through a reduction in output, which tended also to raise unit costs, while unit costs could only be reduced by producing more for sale at a lower price. It was the inability of many papermakers to recognise this basic problem that accounts for the failure of so many firms and attempts at collective

[50] Strenger, 'Bedeutung', pp. 40–2.   [51] Wittigschlager, 'Geschichte', p. 45.
[52] Strenger, 'Bedeutung', pp. 40–4; Wittigschlager, 'Geschichte', pp. 71–6.

action in the latter half of the nineteenth century. A solution to this conundrum came with the realisation that if alternative outlets for the surplus can be found, mass production need not perforce lead to overproduction and the decimation of price. Integral to this strategy are the co-operation of producers and the presence of an international system of trade that permitted almost unlimited access to at least some other nation's markets and protected one's own markets from similar incursions. With the *Verein* and other associations and Great Britain's adherence to free trade in paper such a combination of factors did exist for German producers. As the *Verein* developed and acquired ever larger membership, it became ever more likely and capable of acting upon this realisation. Moreover, in this aim the association was aided by the legal enforceability of its agreements in German law; a development which, with its 358 members in 1910, resolved the association's considerable free-rider problem.[53]

From the beginning of 1876 a lively debate over the industry's woes and the need for a syndicate occupied the pages of the *Wochenblatt für Papierfabrikation*.[54] These pleas, however, were not to be acted upon until the turn of the century when the continued decline of prices and growing competition of Scandinavian and North American producers, not to mention the need to contend with cartels set up by the producers of pulp, coal, and chemicals, proved to tip the balance in favour of cartelisation. When it did begin, the movement was swift and pervasive. Though many of these cartels were not to last much longer than their predecessors, the idea of industry co-operation had taken hold, so that on the demise of one particular arrangement, a new one soon arose to take its place. Writing at the time Demuth appears somewhat amazed at the activity and speed of change going on all around him: 'Not a week goes by in which there is not a wedding of producers and prices fixed. Not a week, in which from one of these a minimum price is not set or a rise announced.'[55]

Between December 1899 and January 1900 a long string of cartels were founded, including *inter alia* those set up by the producers of wallpaper, wood-free paper, packing paper, cardboard, coloured paper, printing paper, writing paper and envelopes. In July of 1900 a brown paper syndicate was established that represented 50 per cent of the production of northern Germany. It subsequently strengthened its hold over the market by coming to an agreement with the major southern

[53] M. John, *Politics and the Law in Late Nineteenth-century Germany: the Origins of the Civil Code* (Oxford, 1989).
[54] For example, see the extracts given in Wittigschlager, 'Geschichte', p. 78.
[55] Demuth, 'Papierfabrikation', p. 207.

German producer of brown paper, Fabrik Teisach, which controlled a number of mills in the region. The terms of the arrangement guaranteed each party to the agreement a virtual monopoly in certain regions of the country. Production for the domestic market, however, was initially to be curtailed by up to 20 per cent. Exports were not affected by this restriction, but were disposed of by the syndicate at prices usually at least 5 per cent below those prevailing domestically. Most other cartels had a similar policy towards exports.[56] By the beginning of the new century so complete had been this process of cartellisation that Demuth could claim that 'the entire industry had merged'.

The most important of these cartels was the *Verband deutscher Druckpapierfabriken* (Association of German printing paper factories), also commonly known just as the Kartell. Its importance stems from the fact that printing paper, constituting 31.5 per cent of the German industry's total output of paper in 1897,[57] was the largest sector of the industry both within Germany and internationally. It was thus the most important commodity in the international paper trade, not just because of its importance as an export, but also because it invariably tended to be the sector of the industry at the forefront of technological and organisational change. Survival in this branch of the trade perhaps to a greater extent than any other was determined by the papermaker's ability to minimise average costs by running his mills to their maximum capacity. Without co-ordination and an equally rapid advance in demand, this inevitably led to overproduction and plunging prices. From its post-war peak of 73.84 marks per 100 kg, the price of newsprint steadily fell to less than a third of what it had been. Perturbed by the trend and its devastating effects on profit margins, the producers of printing paper felt the need for collective action. The Verband was the result.[58]

Formed on 15 October 1900 after three years of negotiations, the Verband consisted of twenty-nine of the most important German papermakers, controlling together about 70 per cent of the entire German output of printing paper. The initial combination lasted until the end of 1905 when it was renewed and eight more mills were admitted.[59] With its headquarters in Berlin, it took the form of a Public

---

[56] *Ibid.*, pp. 249–50. For example, a syndicate of imitation parchment paper producers pursued a similar policy of exporting paper at lower than the domestic price. Cartels also opened up sample rooms and arranged credit for its customers. *PMC*, 10 March 1886, p. 76; 11 October 1896, p. 330; 10 April 1886, p. 124.
[57] Demuth, 'Papierfabrikation', p. 201.
[58] Strenger, 'Bedeutung', p. 20; Wittigschlager, 'Geschichte', p. 44.
[59] Spicer, *Paper Trade*, p. 232.

Limited Company (GmbH), with participating members functioning as partners and bound by the statutes of the association. Its guiding principle and motto was to maintain profitable paper prices: *Papierpreise! Zahlen reden!*[60] It was deemed necessary to raise prices not just for reasons of profit, but so as to ensure the survival of an industry that employed thousands. As one large entrepreneur of the time explained: 'Profit margins per kilo of product have been pressed down to their lowest levels, which is economically unhealthy and in the long run dangerous to the general public.'[61]

The means by which the objective of price control was achieved was through output restrictions. Each member of the Verband was usually assigned a half-yearly production quota which upon completion was delivered exclusively to the cartel's own warehouses. Distribution from there was handled solely by officials of the Verband based in Berlin and Frankfurt-am-Main, who every six months calculated and allocated the profits to members. An array of terms and rules governed the quota system, and any breach of these either in the shape of overproduction or for not meeting the stipulated quality and make-up of the consignment was met with harsh punishment usually in the form of hefty fines. The quota system ensured that excessive quantities of printing paper did not reach the domestic market and force down the price. On some occasions this required firms to reduce their production for the local market by as much as 30 per cent.[62]

Output restriction by itself, however, was not enough, since producers were concerned with lowering costs as well as raising prices. Moreover, the policy of output restriction provoked a backlash of criticism from the media, which succeeded in forcing an inconsequential inquiry by the *Reichsamt des Innern* (Ministry of the Interior) into the alleged malfeasances of the Verband.[63] It also prompted a fruitless attempt on the part of the German newspapers to set up a countervailing purchasing authority for German newspaper publishers (*Einkaufsstelle deutscher Zeitungsverlager*).[64] With such resistance to the use of output restrictions, exports, therefore, were forced to assume a very important part of the Verband's policy. This was hardly a new development. In 1900 12 per cent of the production and 14 per cent of the revenue of the German paper and board industry already came through exports. In this period of cartelisation, however, this grew to over 15 per cent.[65]

[60] Strenger, 'Bedeutung', p. 21.   [61] *Ibid.*, p. 22.
[62] Demuth, 'Papierfabrikation', pp. 245–6; Wittigschlager, 'Geschichte', p. 100.
[63] The pressure, however, did force the Verband to limit its price rises, so as not to undermine support in the Reichstag for the higher import duties it wanted. Wittigschlager, 'Geschichte', p. 97.
[64] Demuth, 'Papierfabrikation', p. 245.   [65] Hülbrock, 'Organisation', p. 168.

Exports were thus exempted from the quota system and members were encouraged to sell their surpluses through the Verband's international selling agency located in Hamburg. This agency disposed of the papermaker's excess production, which was usually greater in harder times, wherever possible, paying the producer a half a pfennig per kilo above the received price.[66] It was an offer eagerly taken up by producers seeking to attain lower costs without sacrificing domestic prices. One consequence of the policy, however, was that the price of German printing paper abroad, as with other paper, was often considerably lower (10 to 15 per cent) than it was at home. In 1902, for example, the price of the 377,200 quintals of paper sold by the Verband in Germany till June of that year was 23.26 Marks per quintal, a price significantly above the 20.09 Marks per quintal it received for the 28,300 quintals it managed to export over the same six months. The same pricing pattern also appeared in the second half of the year when the Verband sold 423,200 quintals in its home market at 22.13 Marks per quintal and 53,200 quintals to foreigners at just 19.82 Marks. Taking these figures as indications of the influence of the Verband, we see that foreign prices were lower than domestic prices in the first six months by 14 per cent and in the second by 10 per cent, or for the whole of 1902 by 12 per cent on average.[67] These type of figures closely match the textbook definition of dumping. Yet to the German producer there was absolutely nothing objectionable in this export policy, for without it it was widely recognised at the time, 'the Syndicate would not have succeeded in maintaining appropriately high prices in the market place'.[68]

A diagrammatic representation

It might prove helpful at this point to model this behaviour, so that we might better comprehend how it was possible, and how the British industry might have responded to it. Such a model is presented in Figures 9.1 and 9.2 which portray the German and British markets for paper.[69] Naturally, Figures 9.1 and 9.2 represent a simplification of a much more complex reality, where *inter alia* there are more than two nations involved,

---

[66] Demuth, 'Papierfabrikation', pp. 245–6. Other cartels adopted this practice. *PTR*, 26 October 1883, p. 161; *PMC*, 10 February 1886, p. 46.

[67] *Abstract of the Proceedings of the German Commission on Kartells*, p. 410; *Report on the German Paper Industry*, p. 178. For another example, see *PMC*, 10 August 1886, pp. 261–2.

[68] Demuth, 'Papierfabrikation', p. 246.

[69] The model resembles in some regards those presented in G. Basevi, 'Domestic demand and ability to export', *Journal of Political Economy* 78 (1970), 330–7, and J. A. Brander and D. A. Krugman, 'A reciprocal dumping model', *Journal of International Economics* 15 (1983), 313–21.

each with their own, often unique, commercial policies; where all paper and board products are not homogeneous; where all pricing and export policies are not carried out with equal efficiency and alacrity, and where change is dynamic rather than static. Nonetheless, this abstraction is useful as a tool to illustrate the essential features.

In the diagrammatic analysis of Figures 9.1 and 9.2 a number of assumptions are made. Firstly, it is assumed in each country that there is a single market for paper and that Britain and Germany produce differentiated products. This may be relaxed without affecting the findings of the analysis by instead assuming that the figures represent one particular sector of the industry, say printing paper, rather than the entire industry. Secondly, it is assumed that the industry is characterised by economies of scale; an assumption borne out by contemporary evidence and discussed in chapter 1. Thirdly, the British market is presumed to be contestable, as also discussed in chapter 1, while the German, dominated by cartels, is monopolistic; and fourthly, trade in paper with and between all other nations, as well as any factors that influence this trade, are assumed to remain constant.

In Figure 9.1 DD and MR represent the total domestic demand and marginal revenue for German paper given British price PB1, while AC and MC are the German industry's unit and marginal cost curve. Similarly in Figure 9.2 dd1 is total British domestic demand for its own paper given German price PD1, while dd2 is the demand at German prices PD2. ac is the British industry's unit cost curve.

Profit-maximising behaviour puts the German industry's equilibrium output at the point at which marginal revenue is equated with marginal cost. The contestability of the British market due to the risk of grade shifting, however, makes this behaviour untenable in Britain, forcing a market equilibrium price equal to average cost.

Our story thus begins with the German industry in equilibrium at PD1 and A output where it is making profits equal to ([PD1–PD2][A-0]), and the British industry at PB1 and X where only normal profits are being made. The German industry then expands its production to B, reducing unit costs to the equivalent of PD3. By exporting (B-A) to Britain the German industry, backed up by cartel agreements, is able to maintain the original domestic equilibrium price of PD1 for the A amount of output marketed in its home market, while it is prepared to receive unit costs (PD3) for the paper exported to Britain. This arrangement affords the German industry above normal profits to the tune of ([PD2–PD3][A-0]).[70]

[70] Of course, if it receives higher than unit cost prices then profits will be even greater. Conversely, German producers may be willing to receive lower than unit cost prices up

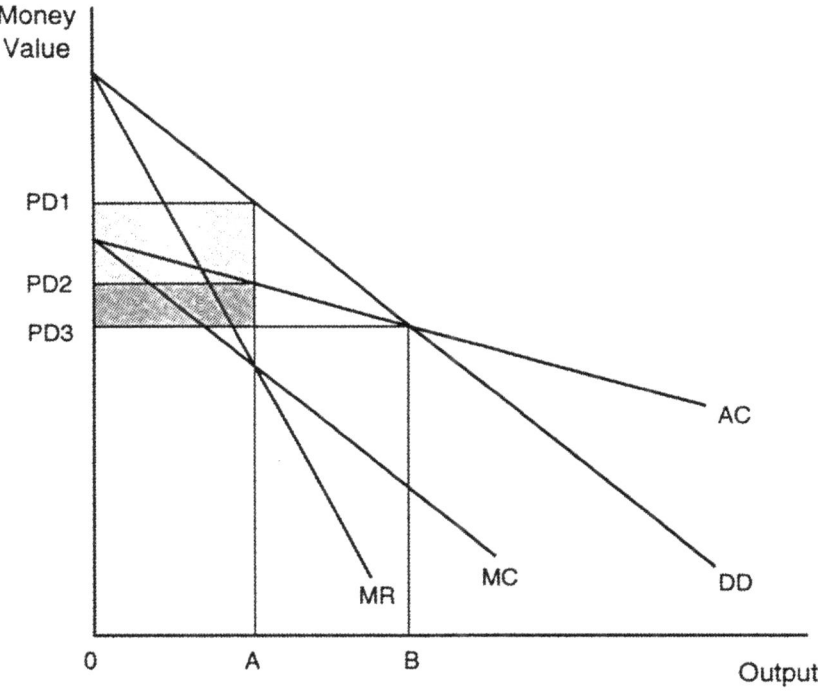

9.1 The German industry

Things look less rosy for the British producer. Cheaper German imports force a substitution away from British products at all price levels. This is represented by the leftward shift in the demand for British paper to dd2. With the domestic price still at PB1 and with this level of demand, there is an excess supply of local paper equal to (X-Z). In time this excess supply forces down British prices to PB2, where all local production is sold, but a loss of ([PB1–PB2] [X-O]) is made.

At this point the British industry has a number of options as to how to react. If it accepts defeat, many firms will go bankrupt and the industry and its output will contract to M, sold at the much higher price of PB*. Moreover, one could expect that further dumping of this kind in Britain would eventually lead to the industry's disappearance. A second response would be to pressure the government to introduce tariffs to protect the industry. This would raise the price of German imports and restore

to the point at which the losses associated with such a price exactly equal the additional profits made in the home market.

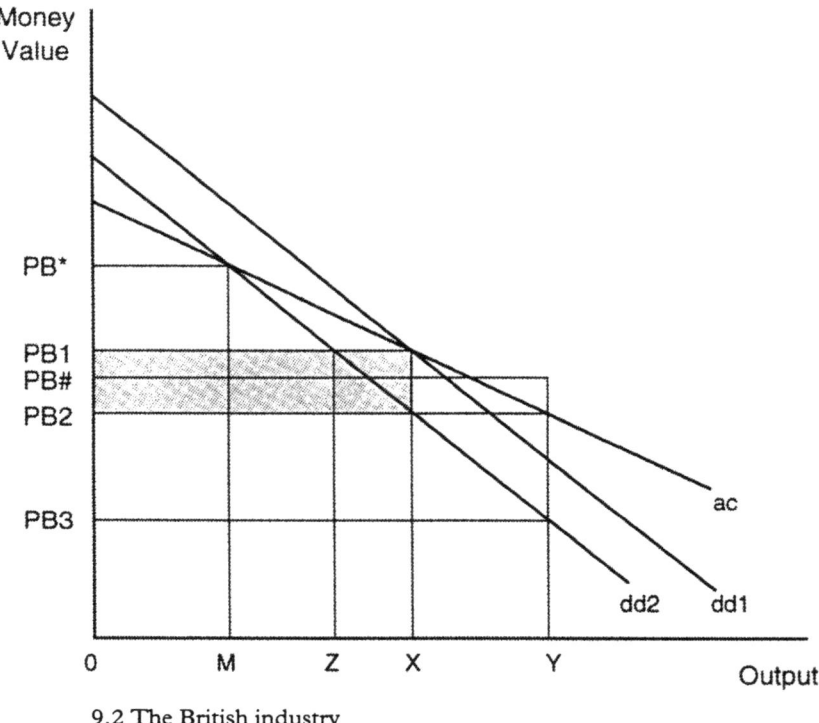

9.2 The British industry

demand for British produce back to the original level or to even better. In the political climate of the day, however, there seemed little hope that this would occur. Alternatively, the British industry could have responded in kind by expanding to Y and exporting (Y-X) to the German market at a price below PD1. If that export price is the same as the domestic (PB2), then the British industry will make normal profits, but if above PB2, such as at PB#, then profits can be made. At PB# these excess profits are equal to ([PB#-PB2][Y-X]). The problem with this strategy is that without the removal of the German tariffs it was impossible. This did not prevent individual British producers from attempting such an approach on the domestic market with devastating consequences. While this may have temporarily improved the prospects of the producer in question, in the end it only fuelled the downward pressure on prices. If the entire industry were to adopt this approach, then according to Figure 9.2, this would mean that Y output would come on to the British market, forcing equilibrium price down to PB3 where losses would be even greater.

Three other possible responses involved strategies not directly related to mass production. The first was to find ways to cut costs other than by increasing scale. Technological and organisational change could shift an industry's unit cost curve inward and give new life to the industry's competitiveness. Likewise, specialisation in products, such as those at the luxury end of the market for which scale was less important and in which Britain had something of a comparative advantage, might shore up the industry's demand. The third possibility was to generate more demand for the home-made paper which would shift the domestic demand curve to the right, or – outside the purview of the model – to find new markets elsewhere for British paper. These were indeed the tactics increasingly adopted by British papermakers.[71]

But what impact did foreign dumping have on the British industry? Ideally our model above needs to be quantified, but given the paucity of reliable data, especially on the supply-side, this is not possible. However, this does not leave us without an idea of the impact. If we assume that all our demand and supply functions are log linear, then it can be shown that the long-run import price elasticity of British paper output and prices would have been:

$$\delta \log Q / \delta \log P_\mathrm{m} = n_\mathrm{m}/1 - \mu n_Q \ (1) \text{ and } \delta \log P / \delta \log P_\mathrm{m} = \mu n_\mathrm{m}/1 - \mu n_Q \ (2)$$
$$\mu = \{(1/\alpha + \beta) - 1\}$$

where Q is the industry's output, P_m and P import and domestic prices, n_m and n_Q the import and own price elasticity of domestic demand, and α and β are the input elasticities of output.[72] This is useful information, as all of these parameters can be directly estimated from available knowledge of the industry.

Of the parameters, n_m is the most troublesome. Employing a number of theoretical assumptions, however, it too can be estimated. If the reciprocity theorem that relates cross price elasticities to relative amounts holds, this then implies that the ratio of cross price elasticities of domestic and import demand can be equated with the ratio of the expenditure on the two types of paper.[73] This permits us to say that if $P_m M/PQ = s$, then

$$s = n_m / \Theta_Q \text{ or } n_m = s(\Theta_Q) \ (3)$$

where M is the amount of imported paper and Θ_Q the cross price

---

[71] See, for example, Bartlett, 'Alexander Pirie and Sons', pp. 30–2.

[72] See J. S. Foreman-Peck, 'Tariff protection and economies of scale: the British motor industry before 1939', *Oxford Economic Papers* 31 (1979), 237–57 where proof and derivation of the techniques employed here are given.

[73] J. R. Hicks, *A Revision of Demand Theory* (Oxford, 1956), pp. 120–9.

elasticity of imported demand. The next assumption is that consumers do not suffer from money illusion, so that the sum of all elasticities for both types of demands is equal to zero, and hence the difference between the import price elasticity of import and domestic demand is equal to the difference between the own price elasticity of domestic and import demand. Subtracting import demand from domestic demand and taking first differences together with our assumption, we are left with

$$\Delta \log M - \Delta \log Q = (\Delta \log P_m - \Delta \log P)(n_Q - \Theta_Q).$$

Rearranging, we get

$$(\Delta \log M - \Delta \log Q)/(\Delta \log P_m - \Delta \log P) = \sigma = n_Q - \Theta_Q$$

or $\Theta_Q = n_Q - \sigma$ where σ is the elasticity of substitution. Substituting into (3) $n_m = s(n_Q - \sigma)$ and then (1) and (2) we get

$\delta \log Q / \delta \log P_m = s(n_Q - \sigma)/1 - \mu n_Q$ and $\delta \log P / \delta \log P_m = \mu s(n_Q - \sigma)/1 - \mu n_Q$.

In the following calculations the values of these parameters have been estimated in a manner designed to determine the minimum likely effect of dumping. In chapter 1, it will be recalled the own price elasticity of domestic demand was estimated as -0.595. The actual value therefore probably did not exceed -0.7. The parameter s is more difficult to ascertain, as it varies positively with changes in P_m. Its value in 1913, for example, was 0.32, but the average value employed in the following analysis was 0.17. The elasticity of substitution used, -8, is the average for the years immediately following 1861 when the tariffs on foreign paper were dropped and when presumably most of the change can be ascribed to the tariff-induced price change. This figure is also not significantly different for that of the whole period (-7.3). Finally, because μ is composed of the output elasticities, its value depends on whether the technology exhibits increasing returns to scale or not. In order to underestimate the effects, constant returns to scale, unless otherwise stated, are assumed, meaning that $\alpha + \beta = 1$ and $\mu = 0$.

On the basis of the above parameter estimates it can be demonstrated that the minimum likely import elasticity of British output was 1.2. Even allowing for considerable increasing returns $(\alpha + \beta = 1.3; \mu = 0.2308)$, the elasticity still only adds up to 1.5. Testimony given at the Tariff Commission indicated that papermakers were generally convinced that an *ad valorem* duty on all foreign-made papers of 7.5 to 10 per cent would have been sufficient to have alleviated the effects of all of the 'unfair' practices they faced.[74] Taking 10 per cent then, as the extreme

[74] Tariff Collection, TC3 1/84, p. 23; TC3 1/86, p. 20; TC3 1/87, p. 13; TC3 1/89, p. 7.

effect of foreign dumping in the British market on P_m, the long-run elasticities calculated above suggest that the imposition of a tariff to remove this impact would have raised British production by between 9 and 15 per cent. This figure, however, is undoubtedly an underestimation. Setting s to its highest level, 0.32, and assuming increasing returns to scale – choices that impose an upward bias on the calculations – the maximum effect of a 10 per cent tariff would have been to augment output by 27.8 per cent. In other words, our calculations suggest that the impact of foreign dumping in Britain probably lowered British domestic production by somewhere between 9 and 28 per cent.

British consumers, however, undoubtedly benefited from the practice; by all accounts paying the lowest prices in Europe for their paper.[75] Unfortunately, precise estimation of its welfare effects or the proposed countervailing duties perforce requires information on consumer and producer surpluses that is just not available. The net effect, however, would almost certainly have been a gain to British society, since virtually all benefited from lower prices, while only the relatively few employed in the trade were directly harmed. Moreover, the lower prices local producers could have guaranteed as a result of the economies of scale made possible by a tariff were unlikely to have altered this situation. Even under very favourable circumstances ($\alpha + \beta = 1.3$), a 10 per cent *ad valorem* tariff on foreign-made paper would have lowered the price of home-made paper at the very most by the 3.4 per cent. Given the huge scale of import penetration (up to half of all paper consumed in some years), this was hardly of a sufficient magnitude to ensure that the average price paid by the British consumer for his or her paper would actually fall with the introduction of such a tariff.

Conclusion

This chapter has focussed on the effects and importance of Britain and Germany's commercial policies on the development of their respective paper industries. Being an industry where cost advantages could be had by expanding output and where the threat of overproduction remained ever a reality, the existence of an open market, where surplus production could be offloaded without affecting domestic prices, was a godsend. This is precisely the situation the combination of British free trade in paper and German protectionism provided for the German producer. Unhindered by import duties, German and other continental and American papermakers were able to find a market for their increased

[75] *Ibid.*, TC3 1/85, p. 15.

production without the worry of retaliation in kind from their British competitors. In addition to reducing unit costs in the short run this type of dumping also paid dividends in the long run by quickening the pace of technological change derived from experience. One weakness of this strategy though was that its benefits could to a large extent be undermined by similar actions on the part of the suppliers of raw materials to the industry. Increasing cartelisation by, and state protection for, the German producers of cellulose, wood pulp, and chemicals, thus meant that despite all their efforts German paper and boardmakers could never totally escape the dilemma of vanishing profit margins. As a consequence this in turn tended to encourage German paper firms to expand further and become even more reliant on the British market. The development of this course of action in Germany was traced in this chapter from its initial practice by individual producers, through its use by the fledgling and variegated trade associations of the second half of the nineteenth century, to its eventual mastering by the massive cartels of the early twentieth century.

In Britain the main effects of dumping were to raise costs and reduce the industry's output by at least 10 per cent. As we saw in chapter 8 it was such reductions in demand that made it hard for the British industry to acquire and replace machinery with the same rapidity as the American which had a large, protected market entirely to itself. Profits also felt the squeeze. Lewis Evans of Dickinsons, for example, saw the firm's profits on newsprint made at its Croxley Mill fall by half following the introduction of the Verband's export policy.[76]

In conclusion, the key implication of German and continental dumping for the study of Britain's decline in importance in the paper trade is that it reduces the share of blame that can be attributed to the industry for the growing importation of foreign-made paper. With protectionism still politically unpalatable in Britain and the foreign keenness to exploit this fact, the dice can be seen as being well and truly loaded against the British paper industry. As a result, the hurdles that the industry had to jump to keep up with its competitors were just that little bit higher and the efforts required from its firms and employees just that little bit greater.

[76] *Ibid.*, TC3 1/84, p. 9.

Conclusion

The decline of Britain as an industrial power is one of the most important, and certainly debated, events of the modern era. Not surprisingly, as in its wake, the face of the international order was irrevocably altered. The aim of the present work is in some small way to make a contribution to the understanding of this pivotal moment in history. It does this by attempting to explain the relative decline in the late Victorian and Edwardian era of the important, but little studied, British paper industry; in particular to determine to what extent the industry's change of fortune at this time can be attributed to its own failings. Over the last nine chapters a wide range of information from various sources has been collated and analysed in an attempt to put together an answer to this question. For the most part the general picture that emerges from this book is one of an industry that up until at least the end of the nineteenth century was performing satisfactorily well in an environment that, it must be said, was becoming increasingly unfavourable to it.

One disadvantageous feature of this environment was the prevailing set of national commercial policies. The combination of Britain's free trade and foreign protectionism actively handicapped the British manufacturer not only by reducing the foreign markets into which he could tap, but also by enabling foreign producers, especially the Germans, to penetrate the British home market with surplus production often sold below cost. Despite the agitation and vehement protestation of many influential papermakers, this detrimental mix of British and foreign commercial policies was never likely to have been altered in this period; at least not without a sea change in British political opinion having first taken place.

Similarly, the transition of the industry to the use of new raw materials not found within the bounds of the kingdom placed local manufacturers at a great disadvantage to their more favourably endowed competitors, while also making the industry ever more dependent on foreign suppliers. The creativeness and vigour of the British papermaker's response to this perceived shortfall in raw materials, however, were more indicative of healthy rather than failing entrepreneurship. Moreover, the ease with which British mills shifted to wood pulp when that material's strengths were fully realised clearly reveals that path dependency need not be permanent and first mover advantages decisive.

Yet it would be wrong to give the impression that the situation was entirely rosy for the British papermaker. After all, market shares were declining, a fact that understandably caused great concern in the trade. Moreover, although British performance in terms of labour productivity growth appears to have been at least comparable to those of most of the other major paper-producing nations of the time, the calculations also show the British industry after 1890 struggling to keep up with the rapid progress of the American industry. Evidence of the transformation that had already occurred in the trade was the fact that by 1900 the cutting edge of papermaking technology had well and truly shifted from Kent and Lancashire to Massachusetts and upstate New York. Once again the restriction of demand for British paper that foreign protectionism assured had a large part to play in this change of technological leadership, but this factor cannot alone account for the magnitude of the divergence experienced. In addition, one also finds evidence of an incipient conservatism developing in the heart of British papermaking, which was beginning to manifest itself more frequently in the British papermakers' disinclination to invest to the same extent as many of their American counterparts in new ideas and strategies beneficial to technological change and productivity growth. In particular, this book has identified the British producer's continued aversion to proactive trade .associations, industry co-operation, trade-related education and technical instruction, as well as productivity linked remunerative schemes, with its poor technological performance *vis-à-vis* the United States from the turn of the century. It was an aversion that seemed to draw its strength from the complacent and conservative business culture that earlier success had bred rather than from a fatal acceptance of the change in comparative advantage that the advent of wood pulp had heralded. After all, in this period only the most prescient of observers could have realised the magnitude of the changes that were to occur. Given that production, exports, machine widths, speeds, and numbers were all still increasing in 1913, the widespread entertainment of such a defeatist idea would have been extremely premature indeed. There were in any case other rational strategies that could have been adopted to counteract the ascendancy of wood pulp; establishing a technological lead in the actual processing and making of paper only being the most obvious. As such then, this debilitating conservatism amongst British papermakers that emerged at the end of the nineteenth century was important as it robbed the industry of one avenue of offsetting, or at least postponing, the effects of the shift in comparative advantage that the appearance of wood pulp on the scene had initiated.

This book also has implications that extend beyond the paper industry. In its investigation of American and British labour organisations it finds an impotence, if not actually a simple lack of intent, in British papermaking unions to influence work practices in the ways that have been frequently adduced to explain the technological divergence between the two countries. This was not because British machine-mill operatives were more accepting of, or indifferent to, the changes that their bosses wanted to implement on the shopfloor, but because their collective voice was weak, divided, and disorganised and in disputes had no craft tradition or custom to call upon for succour and justification. As such, this work casts doubt on those who perceive Britain's economic decline as a direct consequence of the strength and recalcitrance of its workforce, as well as on other simplistic attempts to force all British economic history to lie in the Procrustean bed of a British 'system' of production. Although there may be cases where such generalisations are true, this should not lead us to forget the actual diversity of experience in British industry of the nineteenth century.

Diversity of experience also seems to be central to the Anglo-American labour productivity gap in manufacturing that opened up in the nineteenth century, a fact that tends to make the search for a single explanation for the phenomenon for all industries in all periods very difficult indeed. In the paper industry at least, this study has found that the labour productivity gap had its explanation in a number of factors whose importance varied from period to period. Around the mid-century, for example, Britain's choice of raw material and less frequent use of labour-saving machinery were crucial, while after 1890 America's faster rate of technological accumulation appears to have assumed prominence. In broad terms, however, the explanation for the American paper industry's relatively higher levels of labour productivity put forward in this book supports the theme, recurrent in the literature, of America's relatively expensive labour and greater access to resources and demand acting in conjunction with a degree of Yankee ingenuity. Analysis of the Anglo-American productivity gap also illustrates the fact that technological leadership need not inevitably lead to labour productivity leadership. As we have seen, between 1860 and 1890 British labour productivity was lower than American despite the fact that its paper-machine technology was more advanced.

This work also makes a contribution to the debate on Britain's relative economic decline in its attempt to put the discussion on entrepreneurial failure and technological change in the late Victorian and Edwardian context on a more analytical and theoretical footing. The model of entrepreneurial decision-making it develops, based upon the notion that decision rules, like information, are constantly changing and improving, provides a more accurate way of assessing entrepreneurial activity than has usually been the case. Similarly, the book's exploration of the determinants of the rate of innovation achieved via learning and experience in production in the paper industry adds both theoretical and empirical flesh to recent work on technological accumulation and clearly has relevance to other modern, scale-intensive industries, such as the motor and the iron and steel industries, where the same type of technological change was common. Together with the model of entrepreneurial decision-making outlined in this book, these findings offer fresh new ways of looking at a very old, but important debate.

Bibliography

PRIMARY SOURCES

British Library of Political and Economic Science:
Tariff Commission Collection.
Webb Trade Union Collection.
Bowaters plc: Bowaters plc and Edward Lloyd Limited Archives.
Centre for Kentish Studies, Maidstone:
Edwin Amies Records.
Hollingworth (Turkey Mill) Papers.
Hagley Museum and Library, Delaware, USA:
Pamphlet Collection.
J. Arthur Murphy's 1868 List of Paper Manufacturers.
Thomas H. Savery Collection.
Modern Records Centre, University of Warwick: NUPMW Annual Reports.
Nuffield College, Oxford: Nuffield College Social Reconstruction Survey.
Oxfordshire County Record Office: Cannon and Clapperton Paper Mill Records.
Scottish Records Office, Edinburgh:
James Bertram and Sons Limited Records.
Messrs Brown and Company Limited Records.
Alexander Cowan and Sons Ltd Records.
Scottish Pulp and Fibre Company Limited Records.

PARLIAMENTARY PAPERS AND OFFICIAL PUBLICATIONS

United Kingdom
'Abstract of the Proceedings of the German Commission on Kartells', in The Board of Trade, *Index to the Second Volume of Memoranda, Statistical Tables and Charts prepared in the Board of Trade with Reference to various Matters bearing on British and Foreign Trade and Industrial Conditions* (London: HMSO, 1905).
Annual Statement of the Trade and Navigation of the United Kingdom with Foreign Countries and British Possessions (London: HMSO, 1860–1914).
Board of Trade Report by J. Lowry Whittle on Profit-sharing (London: HMSO, 1890).

Board of Trade Report by D. F. Schloss on Profit-sharing (London: HMSO, 1894).
Board of Trade Report on 'Gain-sharing' and Certain other Systems of Bonus on Production (London: HMSO, 1895).
Board of Trade Report on Profit-sharing and Labour Co-partnership in the United Kingdom (London: HMSO, 1912).
Board of Trade Report on Profit-sharing and Labour Co-partnership Abroad (London: HMSO, 1914).
Comparative Trade Statistics. Statistical Tables showing the Progress of British Trade and Production, 1854–1895 (London: HMSO, 1896).
Copy of a Correspondence between the Departments of the Treasury and Board of Trade, in regard to the increasing scarcity of the materials for the fabrication of paper, *British Parliamentary Papers* LXV (1854).
First Report of the Royal Commission appointed to inquire into the Depression of Trade and Industry, *British Parliamentary Papers* XXIII (1885).
First, Second, and Third Census of Production (London: HMSO, 1907, 1914, 1931).
Fourth Report of the Commissioners on the Employment of Children and Young Persons in Trades and Manufacturers not already regulated by Law, *British Parliamentary Papers* XX (1865).
Report from the Select Committee on Paper (Export Duty on Rags), *British Parliamentary Papers* XI (1861).
Report on the German Paper Industry and Export Trade by Consul-General Sir A. Ward, *British Parliamentary Papers* CXXII (1906).
Report on Japanese Paper-mills by Mr Oswald White, Second Assistant in His Majesty's Consular Service in Japan, *British Parliamentary Papers* LXXXVII (1907).
Report of an Enquiry by the Board of Trade into the Earnings and Hours of Labour of Workpeople of the United Kingdom. vol. VIII – Paper, Printing, etc., Trades; Pottery, Brick, Glass and Chemical Trades; Food, Drink and Tobacco Trade; and Miscellaneous Trades in 1906, *British Parliamentary Papers* CVIII (1912–13).
Report of the Committee on Packing and Wrapping Paper Industries, *British Parliamentary Papers* XV (1924–5).
Return of the Weight of Paper used in all the Government, Revenue and Parliamentary Offices, including printed and plain during the Year 1857, as nearly as can be ascertained, *British Parliamentary Papers* XXXIV (1857–8).
Returns of Factories and Workshops, *British Parliamentary Papers* LXII (1871).
Royal Commission on Labour – Textiles, Clothing, Chemical, Buildings and Miscellaneous Trades (Group C), *British Parliamentary Papers* XXXIV (1893–4).
Second Report of the Royal Commission on Technical Instruction, *British Parliamentary Papers* XXXI (1884).
Second Report of the Royal Commission appointed to inquire into the Depression of Trade and Industry, *British Parliamentary Papers* XXIII (1886).
Statistical Abstract for the Principal and other Foreign Countries (London: HMSO).
Third Report of the Royal Commission appointed to inquire into the Depression of Trade and Industry, *British Parliamentary Papers* XXIII (1886).

Other nations

Bureau of the Census, *Historical Statistics of the United States* (Washington DC: US Government Printing Office, 1976).

Department of Labor, *First Annual Report of the Commissioner of Labor* (Washington DC: US Government Printing Office, 1886).

Eighth to Fifteenth Census of Manufactures (Washington DC: US Government Printing Office, 1850–1914).

Select Committee of the House of Representatives, *Pulp and Paper Investigation Hearings* (Washington DC: US Government Printing Office, 1909).

Statistiches Jahrbuch für das deutsche Reich (Berlin: Verlag von Puttkammer und Mühlbrecht, 1880–1914).

Tariff Board, *Pulp and Newsprint Paper Industry* (Washington DC: US Government Printing Office, 1911).

PERIODICAL PUBLICATIONS

The Board of Trade Journal
Chronological and Descriptive Index of Patents applied for and Patents granted
Chronological Index of Patents of Invention
German Export Review
Paper Maker and British Paper Trade Journal
Paper Maker's Circular and Rag Merchant Gazette and Price Current
Paper Mill Directory of England, Scotland, and Ireland and Year Book of the Paper-making Trade
Paper Mill Directory of the World
Paper Trade Journal
Paper Trade Review
Patents for Inventions. Abridgements of Specifications. Class 96: Paper and Paper-making

SECONDARY SOURCES

Abramovitz, M., 'Catching up, forging ahead and falling behind', *Journal of Economic History* 46 (1986), 385–406.

Ahmad, A., 'On the theory of induced innovation', *Economic Journal* 76 (1966), 344–57.

Aldcroft, D., 'The entrepreneur and the British economy, 1870–1914', *Economic History Review*, 2nd series, 17 (1964), 113–34.

Allen, R. C., 'Entrepreneurship and technical progress in the Northeast Coast pig iron industry: 1850–1913', *Research in Economic History* 6 (1981), 35–71.

'Collective invention', *Journal of Economic Behaviour and Organization* 4 (1983), 1–24.

American Paper and Pulp Association, *The American Paper and Pulp Association: Constitution, Officers, Committees, Members, Historical Sketch*, New York: Lockwood, 1897.

Amigo, E., Neuffer, M., and Maunder, E. R., *Beyond the Adirondacks: The Story of St Regis Paper Company*, Westport, CT: Greenwood Press, 1980.

Armstrong, G. R., *An Economic Study of the New York Pulp and Paper Industry*, Syracuse: State University College of Forestry, 1968.

Arrow, K., 'The economic implications of learning by doing', *Review of Economic Studies* 29 (1962), 155–73.

Arthur, W. B., 'Competing technologies, increasing returns, and lock-in by historical events', *Economic Journal* 99 (1989), 116–31.

Ashworth, W., *An Economic History of England 1870–1939*, London: Methuen, 1960.

Association of Makers of Esparto Papers, *Esparto Paper*, London: Newman Neame, 1956.

Atkinson, A. B. and Stiglitz, J. E., 'A new view of technological change', *Economic Journal* 79 (1969), 573–8.

Balassa, B., 'Trade liberalisation and 'revealed' comparative advantage', *Manchester School* 33 (1965), 99–123.

' 'Revealed' comparative advantage revisited: an analysis of relative export shares of the industrial countries, 1953–71', *Manchester School* 45 (1977), 327–44.

Balderston, C. C., *Profit Sharing for Wage Earners*, New York: J. J. Little and Ives, 1937.

Bartlett, J. N., 'Alexander Pirie & Sons of Aberdeen and the expansion of the British paper industry, c.1860–1914', *Business History* 22 (1980), 18–34.

Basalla, G., *The Evolution of Technology*, Cambridge University Press, 1988.

Basevi, G., 'Domestic demand and ability to export', *Journal of Political Economy* 78 (1970), 330–7.

Baumol, W. J., 'Entrepreneurship in economic theory', *American Economic Review. Papers and Proceedings* 58 (1968), 64–71.

'Productivity growth, convergence and welfare: what the long run data show', *American Economic Review* 76 (1986), 1072–85.

Baumol, W. J., Panzar, J. C. and Willig, R. D., *Contestable Markets and the Theory of Industry Structure*, London: Harcourt Brace Jovanovich, 1982.

Berg, M., (ed.), *Technology and Toil in Nineteenth-century Britain*, London: CSE Books, 1979.

Bettendorf, H. T., *Paperboard and Paperboard Containers: A History*, Camden, NJ: Board Products Publishing Co., 1946.

Blanchflower, D. G. and Oswald, A. J ., 'Profit sharing – can it work?', *Oxford Economic Papers* 39 (1987), 1–19.

Boese, D. L., *Papermakers: The Blandin Paper Company and Grand Rapids, Minnesota*, Grand Rapids: Charles K. Blandin Foundation, 1984.

Böhmert, V., *Die Gewinnbeteiligung. Untersuchungen über Arbeitslohn und Unternehmergewinn*, Leipzig: F. A. Brockhaus, 1878.

Bowie, J. A., *Sharing Profits with Employees*, London: Pitman and Sons, 1922.

Branch, M. L., 'The paper industry in the Lake Regions, 1834–1947', Ph.D. thesis, University of Wisconsin, 1954.

Brander, J. A. and Krugman, D. A., 'A reciprocal dumping model', *Journal of International Economics* 15 (1983), 313–21.

British Productivity Council, *The British Productivity Council Case Studies 2: Plant Maintenance*, London: British Productivity Council, 1956.

Broadberry, S. N., 'Comparative productivity in British and American manufacturing during the nineteenth century', Warwick Economic Research Paper no. 399, 1992.

'Manufacturing and the convergence hypothesis: what the long run data show', paper presented to the ESRC Quantitative Economic History Conference, St Antony's College, Oxford, 1992.

'Technological leadership and productivity leadership in manufacturing since the Industrial Revolution: implications for the convergence debate', *Economic Journal* 104 (1994), 291–302.

Broadberry, S. N. and Crafts, N. F. R., 'Explaining Anglo-American productivity differences in the mid-twentieth century', *Oxford Bulletin of Economics and Statistics* 52 (1990), 375–402.

Brotslaw, I., 'Trade unionism in the pulp and paper industry', Ph.D. thesis, University of Wisconsin, 1964.

Brown, M., *On the Theory and Measurement of Technological Change*, Cambridge University Press, 1966.

Bruland, K., *British Technology and European Industrialization: The Norwegian Textile Industry in the Mid-nineteenth Century*, Cambridge University Press, 1989.

Buchheim, C., *Deutsche Gewerbeexporte nach England in der zweiten Hälfte des 19. Jahrhunderts*, Ostfindern: Scripta Mercaturae Verlag, 1982.

Bundock, C. J., *The Story of the National Union of Printing, Bookbinding and Paper Workers*, Oxford University Press, 1959.

Cain, L. P. and Paterson, D. G., 'Biased technical change, scale and factor substitution in American industry, 1850–1919', *Journal of Economic History* 46 (1986), 153–64.

Calder, L., *The First Hundred Years: Perkins-Goodwin Company*, New York: privately published, 1946.

Calkin, M. W., *Sharing the Profits*, Boston: Ginn and Co., 1888.

Carlson, V., 'Associations and combinations in the American paper industry', Ph.D. thesis, Harvard University, 1931.

Carter, C. F. and Williams, B. R., *Industrial and Technical Progress*, Oxford University Press, 1957.

Casson, M., *The Entrepreneur: An Economic Theory*, Aldershot: Gregg Revivals, 1991.

'Entrepreneurship and business culture', University of Reading Discussion Paper in Economics no. 239, 1991.

'Cultural determinants of economic performance', University of Reading Discussion Papers in Economics no. 258, 1992/3.

Chandler, A. D., 'Development of modern management structure in the US and UK', in L. Hannah (ed.), *Management Strategy and Business Development*, London: Macmillan, 1976, pp. 25–51.

Scale and Scope: The Dynamics of Industrial Capitalism, Cambridge, MA: Belknap Press, 1990.

Chapman, S. J., *Work and Wages*, London: Longmans, Green and Co., 1904.

276   Productivity and performance

Chater, M., *Family Business: A History of Grosvenor Chater, 1690–1977*, St Albens, Herts.: Chater and Co. Ltd., 1977.

Clapham, J. H., *An Economic History of Modern Britain*, Cambridge University Press, 1938.

Clapperton, R. H., *The Paper-making Machine: Its Invention, Evolution and Development*, London: Pergamon Press, 1967.

Clark, V. S., *History of Manufactures in the United States*, New York: McGraw-Hill, 1929.

Cohen, A. J., 'The economic determination of technological change: a theoretical framework and a case study of the US pulp and paper industry, 1915–1940', Ph.D. thesis, Stanford University, 1982.

'Technological change as historical process: the case of the US paper and pulp industry, 1915–1940', *Journal of Economic History* 44 (1984), 775–99.

'Factor substitution and induced innovation in North American kraft pulping, 1914–1940', *Explorations in Economic History* 24 (1987), 197–217.

Cole, A., *Business Enterprise in its Social Setting*, Cambridge, MA: Harvard University Press, 1959.

Coleman, D. C., 'Combinations of capital and of labour in the English paper industry, 1789–1825', *Economica* 21 (1954), 32–53.

'Industrial growth and industrial revolutions', *Economica* 89 (1956), 1–22.

The British Paper Industry, 1495–1860: A Study in Industry Growth, Oxford University Press, 1958.

Coleman, D. C. and MacLeod, C., 'Attitudes and new techniques: British businessmen, 1800–1950', *Economic History Review*, 2nd series, 39 (1986), 588–611.

Cowan, R., 'Nuclear power reactors: a study in technological lock-in', *Journal of Economic History* 50 (1990), 541–68.

Crafts, N. F. R., 'Revealed comparative advantage in manufacturing, 1899–1950', *Journal of European Economic History* 18 (1989), 127–37.

Crafts, N. F. R. and Thomas, M. F., 'Comparative advantage in UK manufacturing trade, 1910–1935', *Economic Journal* 96 (1986), 629–45.

Crawford, F. J., 'Manufacturing in the United States, 1860–1870', Ph.D. thesis, University of Wisconsin, 1922.

Crouzet, F., *The Victorian Economy*, London: Routledge, 1982.

Cubbin, J., 'Apparent collusion and conjectural variations in differentiated oligopoly', *International Journal of Industrial Organization* 1 (1983), 155–63.

Cummins, J. G., 'Concentration and mergers in the pulp and paper industries of the United States and Canada, 1895–1955', Ph.D. thesis, Johns Hopkins University, 1960.

David, P. A., *Technical Choice, Innovation, and Economic Growth: Essays on American and British Experience in the Nineteenth Century*, Cambridge University Press, 1975.

'Clio and the economics of QWERTY', *American Economic Review* 75 (1985), 332–7.

Deane, P. and Cole, W. A., *British Economic Growth*, Cambridge University Press, 1967.

Demuth, F., 'Die Papierfabrikation', *Schriften des Vereins für Sozialpolitik* 107 (1903), 197–250.

Dixit, A., 'Investment and hysteresis', *Journal of Economic Perspectives* 6 (1992), 107–32.

Dorwick, S. and Nguyen, D. T., 'O.E.C.D. comparative economic growth 1950–1985: catch up and convergence', *American Economic Review* 79 (1989), 1010–30.

Dosi, G., 'Technological paradigms and technological trajectories: a suggested interpretation of the determinants and direction of technical change', *Research Policy* 11 (1982), 147–62.

Technical Change and Industrial Transformation, London: Macmillan, 1984.

'The nature of the innovative process', in G. Dosi, C. Freeman, R. R. Nelson, G. Silverberg, and L. Soete (eds.), *Technical Change and Economic Theory*, London: Pinter, 1988, pp. 221–38.

Duly, S. J. (ed.), *Timber and Timber Products, including Paper-making Materials*, London: Ernst Bevin, 1924.

Dunbar, J., *Notes on the Manufacture of Wood Pulp and Wood-pulp Papers*, Leith: MacKenzie and Storry, 1894.

Dyson, W., 'The Paper Trade', in *Mosely Industrial Commission to the United States of America, Oct.–Dec. 1902. Reports of the Delegates*, Manchester: Co-operative Printing Society, 1903, pp. 215–19.

Elbaum, B. and Lazonick, W. (eds.), *The Decline of the British Economy*, Oxford: Clarendon Press, 1986.

Elster, J., *Explaining Technical Change*, Cambridge University Press, 1983.

Estrin, S., Grout, P. and Wadhwani, S., 'Profit-sharing and employee share ownership: an assessment', *Economic Policy* 4 (1987), 13–52.

Evans, J., *The Endless Web: John Dickinson & Co. Ltd., 1804–1954*, London: Jonathan Cape, 1955.

Evans, L., *The Firm of John Dickinson and Company Limited*, London: Chiswick Press, 1896.

Fabricant, S., *The Output of Manufacturing Industries, 1899–1937*, New York: NBER, 1940.

Feinstein, C. H., *National Income, Expenditure and Output of the United Kingdom, 1855–1965*, Cambridge University Press, 1972.

Fellner, W., 'Two propositions in the theory of induced innovations', *Economic Journal* 71 (1961), 305–8.

Foreman-Peck, J. S., 'Tariff protection and economies of scale: the British motor industry before 1939', *Oxford Economic Papers* 31 (1979), 237–57.

Frankel, M., *British and American Manufacturing Productivity*, Urbana: University of Illinois Press, 1957.

Frickey, E., *Production in the United States, 1860–1914*, Cambridge, MA: Harvard University Press, 1947.

Friedman, M. and Schwarz, A., *A Monetary History of the United States*, Princeton University Press, 1963.

Geuenich, J., *Geschichte der Papierindustrie im Düren-Jürlicher Wirtschaftsraum*, Düren: Dürener Druckerei und Verlag Carl Hamel, 1959.

Gilman, N. R., *Profit-sharing between Employer and Employee: A Study in the Evolution of the Wages System*, New York: Houghton, Mifflin, and Co., 1889.

Green, C. M., *Holyoke Massachusetts: A Case History of the Industrial Revolution in America*, New Haven: Yale University Press, 1939.

278   Productivity and performance

Guthrie, J. A., *The Newsprint Industry: An Economic Analysis*, Cambridge, MA: Harvard University Press, 1941.

The Economics of Pulp and Paper, Pullman, WA: The State College of Washington Press, 1950.

Habakkuk, H. J., *American and British Technology in the Nineteenth Century*, Cambridge University Press, 1962.

Hamburg, M., *Basic Statistics*, London: Harcourt Brace Jovanovich, 1985.

Hancock, H. B. and Wilkinson, N. B., 'Joshua Gilpin: an American manufacturer in England and Wales. Part II', *Newcomen Society for the Study of the History of Engineering and Technology Transactions* 39 (1960–1), 57–66.

Harley, C. K., 'The shift from sailing ships to steamships, 1850–1890: a study in technological change and its diffusion', in D. N. McCloskey (ed.), *Essays on a Mature Economy: Britain after 1840*, London: Methuen, 1971, pp. 215–34.

Harrison, R. and Zeitlin, J., *Divisions of Labour, Skilled Workers and Technological Change in Nineteenth-century England*, Brighton: Harvester Press, 1985.

Hatton, T. J., 'Profit sharing in British industry, 1865–1913', *International Journal of Industrial Organization* 6 (1988), 69–90.

Hèbert, R. and Link, A., *The Entrepreneur*, New York: Praeger, 1982.

Heinel, Ing., 'Die Maschinen-Industrie', in *Deutschland unter Kaiser Wilhelm II*, Berlin: Verlag von Reimar Hobbing, 1914.

Helps to Profitable Paper Making, Chicago: The *Paper Trade Journal*, 1898.

Herring, R., *Paper and Paper Making, Ancient and Modern*, London: Longmans, 1863.

Hicks, J. R., *Theory of Wages*, Oxford University Press, 1932.

A Revision of Demand Theory, Oxford: Clarendon Press, 1956.

Hills, R. L., *Papermaking in Britain 1488–1988*, London: Athlone, 1988.

Hobsbawm, E. J., *Labouring Men: Studies in the History of Labour*, London: Weidenfeld and Nicolson, 1964.

Hobson, J., *Incentives in the New Industrial Order*, New York: Thomas Setzler, 1923.

Hoffmann, W. G., *British Industry, 1700–1950*, Oxford: Blackwell, 1955.

Das Wachstum der deutschen Wirtschaft seit der Mitte des 19. Jahrhunderts, Berlin: Springer-Verlag, 1965.

Hülbrock, A., 'Organisation und Preisgestaltung auf dem deutschen Papiermarkt unter besonderer Berücksichtigung der Gegenwart', Ph.D. thesis, University of Frankfurt-am-Main, 1927.

Hunter, D., *Papermaking: The History and Technique of an Ancient Craft*, New York: Dover, 1974.

Hutchins, B. L., 'The employment of women in paper-mills', *Economic Journal* 14 (1904), 235–48.

International Correspondence Schools Reference Library, *Sulphuric Acid, Alkalis and Hydrochloric Acid. Manufacture of Paper*, Scranton: International Textbook Co., 1902.

James, J. and Skinner, J. S., 'The resolution of the labor-scarcity paradox', *Journal of Economic History* 45 (1985), 513–40.

Jewkes, J., Sawers, D. and Stillerman, R., *The Sources of Invention*, London: Macmillan, 1969 [1958].

John, M., *Politics and the Law in Late Nineteenth-century Germany: the Origins of the Civil Code*, Oxford: Clarendon Press, 1989.
Johnson, H., 'A new view of the infant industry argument', in I. McDougall and R. Snape (eds.), *Studies in International Economics: Monash Conference Papers*, Amsterdam: New Holland, 1970, pp. 59–76.
Karges, S. B., 'David Clark Everest and Marathon Paper Mills Company: a study of a Wisconsin entrepreneur, 1909–1931', Ph.D. thesis, University of Wisconsin, 1968.
Kay, J., *Paper, its History*, London: Smith, Kay and Co., 1893.
Keir, M., *Manufacturing*, New York: Ronald Press Co., 1928.
Ketelbey, C. D. M., *Tullis Russell: The History of R. Tullis and Company and Tullis Russell & Co. Ltd. 1809–1959*, Markinch, Fife: Tullis Russell & Co. Ltd., 1967.
Kirchner, E., *Die Papierfabrikation in den Ländern der Sektion X der Papiermacher-berufgenossenschaft: historisch-technologische Skizze*, Württemberg: Güntter-Staib, 1911.
Kirzner, I. M., *Competition and Entrepreneurship*, University of Chicago Press, 1973.
'The primacy of entrepreneurial discovery', in I. M. Kirzner (ed.), *The Prime Mover of Progress: The Entrepreneur in Capitalism and Socialism*, London: IRA, 1980, pp. 5–25.
Discovery, Capitalism, and Distributive Justice, Oxford: Blackwell, 1989.
Knight, F. H., *Risk, Uncertainty and Profit*, Boston: Houghton Mifflin, 1921.
Lamoreaux, N. R., 'Industrial concentration and market behavior: the great merger movement in American industry', Ph.D. thesis, Johns Hopkins University, 1979.
Landes, D. S., *The Unbound Prometheus*, Cambridge University Press, 1969.
Lawson, W. K., *British Railways: A Financial and Commercial Survey*, London: Constable & Co. Ltd., 1913.
Lazonick, W., 'Industrial relations and technical change: the case of the self-acting mule', *Cambridge Journal of Economics* 3 (1979), 231–62.
'Competition, specialisation and industrial decline', *Journal of Economic History* 41 (1981), 31–8.
'Production relations, labour productivity, and the choice of technique: British and US cotton spinning', *Journal of Economic History* 41 (1983), 491–516.
Lee, C. H., 'Regional growth and structural change in Victorian Britain', *Economic History Review*, 2nd series, 34 (1981), 438–52.
Leibenstein, H., 'Entrepreneurship and development', *American Economic Review. Papers and Proceedings* 48 (1968), 72–83.
Levin, R. C., Cohen, W. M., and Mowery, D. C., 'R&D appropriability, opportunity, and market structure: new evidence on some Schumpeterian hypotheses', *American Economic Review. Papers and Proceedings* 75 (1985), 20–4.
Levine, A., *Industrial Retardation in Britain, 1880–1914*, London: Weidenfeld and Nicholson, 1967.
Lewchuck, W., *American Technology and the British Vehicle Industry*, Cambridge University Press, 1987.

Lewis, P. W., *A Numerical Approach to the Location of Industry*, University of Hull Press, 1969.

Lindert, P. H. and Trace, K., 'Yardsticks for Victorian entrepreneurs', in D. N. McCloskey (ed.), *Essays on a Mature Economy: Britain after 1840*, London: Methuen, 1971, pp. 239–74.

Machlup, F., *Knowledge and Knowledge Production*, Princeton University Press, 1980.

MacLeod, C., 'Strategies for innovation: the diffusion of new technology in nineteenth-century British industry', *Economic History Review*, 2nd series, 45 (1992), 285–307.

Maddison, A., *Dynamic Forces in Capitalist Development: A Long-run Comparative View*, Oxford University Press, 1991.

Magee, J. F., 'Arthur D. Little, Inc: at the moving frontier', paper presented to the Newcomen Society of the United States, 1986.

Markovitch, T. J., 'L'industrie française de 1789 à 1964 – sources et méthodes', *Cahiers de L'ISEA* AF 4 (1965).

Marshall, A., *Industry and Trade*, London: Macmillan, 1923.

McCloskey, D. N., *Economic Maturity and Entrepreneurial Decline*, Cambridge, MA: Harvard University Press, 1973.

McCloskey, D. N. and Sandberg, L. G. , 'From damnation to redemption: judgements on the late Victorian entrepreneur', in D. N. McCloskey (ed.), *Enterprise and Trade in Victorian Britain*, London: Allen & Unwin, 1981, pp. 55–72.

McGaw, J. A., 'Technological change and women's work: mechanization in the Berkshire paper industry, 1820–1855', in M. M. Trescott (ed.), *Dynamos and Virgins Revisited: Women and Technological Change in History*, Metuchen, NY: Scarecrow Press Inc., 1979, pp. 81–98.

'Accounting for innovation: technological change and business practice in the Berkshire county paper industry', *Technology and Culture* 26 (1985), 703–25.

Most Wonderful Machine: Mechanization and Social Change in Berkshire Paper Making, 1801–1885, Princeton University Press, 1987.

McPherson, J. M., *Battle Cry of Freedom: The Civil War Era*, New York: Ballantine, 1988.

McQuillen, M. J. and Garvey, W. P., *The Best Known Name in Paper: Hammermill. A History of the Company*, Erie, PA: privately published, 1985.

Mises, L. von, *Human Action: A Treatise on Economics*, New Haven: Yale University Press, 1949.

Profit and Losses, South Holland, IL: Consumers–Producers Economic Service, 1951.

Mitchell, B. R., *British Historical Statistics*, Cambridge University Press, 1988.

Muir, A., *The British Paper and Board Makers Association, 1872–1972: A Centenary History*, London: privately printed, 1972.

Mulhall, M. G., *The Progress of the World in Arts, Agriculture, Commerce, Manufactures, Industries, Railways, and Public Wealth*, London: Edward Stanford, 1880.

Industries and Wealth of Nations, London: Longmans, Green and Co., 1896.

Dictionary of Statistics, London: Routledge and Son, 1899.

Munsell, J., *Chronology of the Origin and Progress of Paper and Paper Making*, New York: Garland Publishing, Inc., 1980 [1876].

National Civic Federation, *Profit-sharing by American Employers*, New York: National Civic Federation, 1920.

National Industrial Conference Board, *Practical Experience with Profit-sharing in Industrial Establishments*, Boston: National Industrial Conference Board, 1920.

Nelson, R. R., *Understanding Technical Change as an Evolutionary Process*, Amsterdam: New Holland, 1987.

Nelson, R. R. and Winter, S., *An Evolutionary Theory of Economic Change*, Cambridge, MA: Belknap, 1982.

Nelson, R. R. and Wright, G., 'The rise and fall of American technological leadership: the postwar era in historical perspective', *Journal of Economic Literature* 30 (1992), 1931–64.

OEEC, *The Pulp and Paper Industry in the USA*, Paris: OEEC, 1951.

Ohanian, N. K., *American Pulp and Paper Industry, 1900–1940: Mill Survival, Firm Structure and Industry Relocation*, Westport, CT: Greenwood, 1993.

Page, W. (ed.), *Commerce and Industry*, London: Constable and Co., 1919.

Paper Mill Directory (the editor) of, *The Art of Papermaking: a Guide to the Theory and Practice of the Manufacture of Paper. Being a Compilation from the Best-known French, German, and American Writers*, London: Kent and Co., 1876.

Pavitt, K., 'R & D, patenting and innovative activities: a statistical exploration', *Research Policy* 11 (1983), 33–51.

'Patent statistics as indicators of innovative activities: possibilities and problems', *Scientometrics* 7 (1985), 77–99.

'International patterns of technological accumulation', in N. Hood and I. Vahlne (eds.), *Strategies in Global Competition*, London: Croom Helm, 1988, pp. 126–57.

Persson, K. G., *Pre-industrial Growth: Social Organization and Technological Progress in Europe*, Oxford University Press, 1982.

Phelps Brown, E. H. and Ozga, S. A., 'Economic growth and the price level', *Economic Journal* 65 (1955), 1–18.

Pindyck, R. S., 'Irreversibility, uncertainty, and investment', *Journal of Economic Literature* 29 (1991), 1110–48.

Pollard, S., *Britain's Prime and Britain's Decline*, London: Arnold, 1989.

Post Office Directory of Stationers, Printers, Booksellers, Publishers, and Paper Makers of England, Scotland, Wales and the Principal Towns in Ireland, London: Kelly and Co., 1872.

Progress of Paper, New York: Lockwood, 1947.

Proteaux, A., *Practical Guide for the Manufacture of Paper and Board*, Philadelphia: Henry Carey Bairds, 1866.

Rahmeyer, F., 'The evolutionary approach to innovation activity', *Journal of Institutional and Theoretical Economics* 145 (1989), 275–97.

Rapping, L., 'Learning and World War Two production functions', *Review of Economics and Statistics* 47 (1965), 81–6.

Rawson, H. G., *Profit-sharing Precedents*, London: Stevens and Sons, 1891.

Reader, W., *Bowaters: A History*, Cambridge University Press, 1981.

Rosenberg, N., *Technology and American Economic Growth*, New York: Harper and Row, 1972.

Perspectives on Technology, Cambridge University Press, 1976.

Inside the Black Box: Technology and Economics, Cambridge University Press, 1982.

Rosenberg, N. (ed.), *The American System of Manufacturing*, Edinburgh University Press, 1969.

Rostas, L., *Comparative Productivity in British and American Industry*, Cambridge: NIESR, 1948.

Rothbarth, E., 'Cause of the superior efficiency of the USA industry as compared with British industry', *Economic Journal* 56 (1946), 383–90.

Ruttan, V. and Hayami, Y., *Agricultural Development: An International Perspective*, Baltimore: Johns Hopkins University Press, 1971.

Sahal, D., 'Technological guideposts and innovative avenues', *Research Policy* 14 (1985), 61–82.

Salter, W. E. G., *Productivity and Technical Change*, Cambridge University Press, 1960.

Sandberg, L. G., *Lancashire in Decline*, Columbus, OH: Ohio State University, 1974.

'The entrepreneur and technological change', in R. Floud and D. N. McCloskey (eds.), *The Economic History of Britain since 1700*, Cambridge University Press, 1981, vol. 2, pp. 99–120.

Scherer, F. M., *Industrial Market Structure and Economic Performance*, Chicago: Rand McNally, 1970.

Schmookler, J., *Invention and Economic Growth*, Cambridge, MA: Harvard University Press, 1966.

Schumpeter, J. A., *The Theory of Economic Development*, Oxford University Press, 1961.

S. D. Warren Company, *A History of S.D. Warren Company, 1854–1954*, Westbrook, Maine: Anthoensen Press, 1955.

Shears, W. S., *William Nash of St Paul's Cray Papermakers*, Leeds: Knight and Forster, 1967.

Shorter, A. H., *Papermaking in the British Isles: An Historical and Geographical Study*, Newton Abbot, Devon: David and Charles Ltd., 1971.

Shryock, G. A., *History of the Origin and Manufacture of Straw and Wood Paper*, Philadelphia: American Philosophical Society, 1866.

Simon, H., *Models of Bounded Rationality*, Cambridge, MA: Harvard University Press, 1982.

Sindall, R. W., *Bamboo for Papermaking*, London: Marchant & Singer Co., 1909.

Smith, D. C., 'Wood pulp and newspapers, 1867–1900', *Business History Review* 38 (1964), 328–45.

History of Papermaking in the United States, 1691–1969, New York: Lockwood Publishing Co., 1970.

Smith, J. E. A., *A History of Paper: Its Genesis and its Revelation*, Holyoke, MA: Clark W. Bryan, 1882.

Society for the Encouragement of Arts, Manufactures, and Commerce, *Report of the Committee on the Deterioration of Paper*, London: William Trounce, 1898.

Sokoloff, K. L., 'Inventive activity in early industrial America: evidence from patent records, 1790–1846', *Journal of Economic History* 48 (1988), 813–50.

Sokoloff, K. L. and Khan, B. Z., 'The democratization of invention during early industrialization: evidence from the United States, 1790–1846', *Journal of Economic History* 50 (1990), 363–78.

Solow, R. M., 'Investment and technical progress', in K. J. Arrow, S. Karlin, and P. Suppes (eds.), *Mathematical Methods in the Social Sciences 1959*, Stanford University Press, 1960, pp. 89–104.

Spicer, A. D., 'The paper trade', in H. Cox (ed.), *British Industries under Free Trade*, London: T. Fisher and Unwin, 1904, pp. 201–13.

The Paper Trade, London: Methuen, 1907.

Stevenson, L. T., *The Background and Economics of American Papermaking*, New York: Harper and Brothers Publishers, 1940.

Stigler, G. J., *The Organization of Industry*, Homewood, IL: Richard D. Irving, 1968.

Stiglitz, J. E., 'Learning to learn, localized learning and technological progress', in P. Dasgupta and P. Stoneman (eds.), *Economic Policy and Technological Performance*, Cambridge University Press, 1987, pp. 125–53.

Strenger, F., 'Die wirtshaftliche und soziale Bedeutung der deutschen Papier Herstellenden Industrie im Rahmen der deutschen Volkswirtschaft', Ph.D. thesis, Leopold-Franzens University of Innsbruck, 1930.

Sullivan, R. J., 'The revolution of ideas: widespread patenting and invention during the English Industrial Revolution', *Journal of Economic History* 50 (1990), 349–62.

Sweden as Producer of Wood Goods, Pulp, Paper, Tar and other Forest Products, Stockholm: A. B. Svenska Teknologföreningens Förlag, 1920.

Tariff Commission, the Report of, *The Engineering Industries*, London: P. S. King and Son, 1909, vol. 4.

Temin, P., 'The relative decline of the British steel industry, 1880–1913', in H. Rosovsky (ed.), *Industrialization in Two Systems*, New York: Wiley, 1966, pp. 140–55.

Thompson, A. G., *The Scottish Paper Industry till 1860*, Edinburgh: Scottish Academic Press, 1981.

Tillmans, M., *Bridge Hall Mills: Three Centuries of Paper and Cellulose Film Manufacture*, Tilsbury, Wilts: Compton Press, 1978.

Toutain, J. C., 'La population de la France de 1700 à 1959', *Cahiers de L'ISEA* AF 3 (1963).

Turvey, R., 'Economic growth and domestic rubbish, London 1855–1926', paper presented at the ESRC Conference on Quantitative Economic History at St Antony's College, Oxford, September 1992.

Vandermeulen, D. C., 'Technological change in the paper industry: the introduction of the sulphite process', Ph.D. thesis, Harvard University, 1947.

Veblen, T. *Imperial Germany and the Industrial Revolution*, New York: Macmillan, 1915.

Voelker, K. E., 'The history of the International Brotherhood of Pulp, Sulphite and Paper Mill Workers from 1906 to 1929: a case study of industrial unionism before the Great Depression', Ph.D. thesis, University of Wisconsin, 1969.

Von Tunzelmann, G. N., 'Technical progress during the Industrial Revolution', in R. Floud and D. N. McCloskey (eds.), *The Economic History of Britain since 1700*, Cambridge University Press, 1981, vol. 1, pp. 143–63.

Wadsworth, A. P., 'Newspaper circulation, 1800–1954', paper presented to the Manchester Statistical Society, 9 March 1955.

Watson, N., *The Last Mill on the Esk: 150 Years of Papermaking*, Edinburgh: Scottish Academic Press, 1987.

Weatherill, L., *One Hundred Years of Papermaking: An Illustrated History of Guard Bridge Paper Company Ltd., 1873–1973*, Edinburgh: Constable Ltd., 1974.

Weeks, L .H., *A History of Paper Manufacturing in the United States, 1690–1916*, New York: Burt Franklin, 1969 [1916].

Weston, H. E., *A Chronology of Papermaking in the United States*, Wilmington, DE: Hercules Powder Co., 1945.

Wheelwright, W. B. and Kean, S., *The Lengthened Shadow of One Man*, Fitchburg, MA: privately published, 1957.

Wicken, O., 'Learning, inventions and innovations: productivity increase and new technology in an industrial firm', *Scandinavian Economic History Review* 33 (1985), 144–72.

Wiener, M. J., *English Culture and the Decline of the Industrial Spirit, 1850–1980*, London: Penguin, 1985.

Williams, E. E., *Made in Germany*, London: W. Heinemann, 1896.

Wittigschlager, K., 'Geschichte der Maschinepapierfabrikation in Bayern rechts des Rheins bis 1914', Ph.D. thesis, University of Frankfurt-am-Main, 1930.

Wolff, E. N., 'Capital formation and productivity convergence over the long term', *American Economic Review* 81 (1991), 565–79.

Wray, M., *The British Paper Industry: A Study in Structural and Technological Change*, London: The British Paper and Board Industry Federation, 1979.

Wright, H. E., *Three Hundred Years of American Papermaking*, Washington DC: Smithsonian Institute, 1991.

Index

288   Index

For EU product safety concerns, contact us at Calle de José Abascal, 56–1°,
28003 Madrid, Spain or eugpsr@cambridge.org.